THE DEMON IN THE MACHINE

'This is one of the m⸺
A demonically brillian⸺ that
only the very latest sci⸺ we
have a view from the⸺
Superb' Robyn⸺

'Davies has written ⸺ imaginative book'
Clive C⸺on, *Financial Times*

'Not only does the theoretical physicist, writer and broadcaster set
out to answer the question of what life is, but he does so on his
scientific terms ... Davies takes the reader on a journey down
a mind-blowing rabbit hole of physics, mathematics and biology'
Bianca Nogrady, *Sydney Morning Herald*

'Davies's lucid writing on this emerging scientific area is just what
the pop-sci reader ordered. He is the perfect host to this admittedly
dizzying journey, as he spins yarns of quantum demons, double-headed
worms and everything in-between' *Physics World*

'Paul Davies narrates a gripping new drama in science, in which the
plot is the story of life and the leading actor is information ... If you
want to understand how the concept of life is changing, read this'
Professor Andrew Briggs, University of Oxford

'A fascinating tour of what is known about what life is'
David Deutsch, author of *The Beginning of Infinity*

'Paul Davies is a courageous explorer of the boundaries of what we
can know about our world. This book makes his explorations
available to all who enjoy pushing those boundaries. Written
with a light entertaining touch, even the most abstruse science
acquires the clarity of exposition for which the author is justly
renowned' Denis Noble, University of Oxford, author of
Dance to the Tune of Life: Biological Relativity

'In this characteristically clearly written and engaging book, ranging
from physics to biology and evolutionary theory to neuroscience,
Paul Davies strongly makes the case that, at its core, life is about
information flows. There is much food for thought here. Highly
recommended' George F. R. Ellis, University of Cape Town,
co-author of *The Large Scale Structure of Space-Time*

ABOUT THE AUTHOR

Paul Davies is a Regents' Professor of Physics and Director of the Beyond Center for Fundamental Concepts in Science at Arizona State University. The author of some thirty books, his many awards include the Templeton Prize and the Faraday Prize of The Royal Society. He is a Member of the Order of Australia and has an asteroid named after him.

PAUL DAVIES

The Demon in the Machine

*How hidden webs of information
are solving the mystery of life*

PENGUIN BOOKS

PENGUIN BOOKS

UK | USA | Canada | Ireland | Australia
India | New Zealand | South Africa

Penguin Books is part of the Penguin Random House group of companies
whose addresses can be found at global.penguinrandomhouse.com

First published by Allen Lane 2019
Published in Penguin Books 2020

005

Copyright © Paul Davies, 2019

The moral right of the author has been asserted

Typeset by Jouve (UK), Milton Keynes
Printed and bound in Great Britain by Clays Ltd, Elcograf S.p.A.

A CIP catalogue record for this book is available from the British Library

ISBN: 978-0-141-98640-1

'The curiosity remains ... to grasp more clearly how the same matter, which in physics and in chemistry displays orderly and reproducible and relatively simple properties, arranges itself in the most astounding fashions as soon as it is drawn into the orbit of the living organism. The closer one looks at these performances of matter in living organisms the more impressive the show becomes. The meanest living cell becomes a magic puzzle box full of elaborate and changing molecules ...'

– *Max Delbrück**

* Max Delbrück, *Transactions of the Connecticut Academy of Arts and Sciences*, vol. 38, 173–90 (December 1949)

Contents

Preface   I

1. What *is* Life?   5
2. Enter the Demon   27
3. The Logic of Life   67
4. Darwinism 2.0   109
5. Spooky Life and Quantum Demons   144
6. Almost a Miracle   166
7. The Ghost in the Machine   184
   Epilogue   209

   Further Reading   219
   Notes   225
   Illustration Credits   236
   Index   237

Preface

There are many books about what life does. This is a book about what life *is*. I'm fascinated by what makes organisms tick, what enables living matter to do such astounding things – things beyond the reach of non-living matter. Where does the difference come from? Even a humble bacterium accomplishes things so amazing, so dazzling, that no human engineer can match it. Life looks like magic, its secrets cloaked by a shroud of impenetrable complexity. Huge advances in biology over the past decades have served only to deepen the mystery. What gives living things that enigmatic *oomph* that sets them apart from other physical systems as remarkable and special? And where did all this specialness come from in the first place?

That's a lot of questions – big questions too. I've been preoccupied with them for much of my working life. I'm not a biologist, I'm a physicist and cosmologist, so my approach to tackling big questions is to dodge most of the technicalities and home in on the basic principles. And that's what I do in this book. I've tried to focus on the puzzles and concepts that really matter in an attempt to answer the burning question: *what is life?* I am by no means the first physicist to ask it; I take as my starting point a series of famous lectures entitled 'What is Life?' by the great quantum physicist Erwin Schrödinger, delivered three generations ago, addressing a question that Darwin dodged. However, I think we are now on the threshold of answering Schrödinger's question, and the answer will usher in a whole new era of science.

The huge gulf that separates physics and biology – the realm of atoms and molecules from that of living organisms – is unbridgeable without fundamentally new concepts. Living organisms have goals and purposes – the product of billions of years of evolution – whereas atoms and molecules just blindly follow physical laws. Yet somehow the one has to come out of the other. Although the need to reconceptualize life as a *physical* phenomenon is widely acknowledged in the scientific community, scientists frequently downplay how challenging a full understanding of the nature and origin of life has proved to be.

The search for a 'missing link' that can join non-life and life in a unitary framework has led to an entirely new scientific field at the interface of biology, physics, computing and mathematics. It is a field ripe with promise not only for finally explaining life but in opening the way to applications that will transform nanotechnology and lead to sweeping advances in medicine. The unifying concept that underlies this transformation is *information*, not in its prosaic everyday sense but as an abstract quantity which, like energy, has the ability to animate matter. Patterns of information flow can literally take on a life of their own, surging through cells, swirling around brains and networking across ecosystems and societies, displaying their own systematic dynamics. It is from this rich and complex ferment of information that the concept of agency emerges, with its links to consciousness, free will and other vexing puzzles. It is here, in the way living systems arrange information into organized patterns, that the distinctive order of life emerges from the chaos of the molecular realm.

Scientists are just beginning to understand the power of information as a *cause* that can actually make a difference in the world. Very recently, laws that interweave information, energy, heat and work have been applied to living organisms, from the level of DNA, through cellular mechanisms, up to neuroscience and social organization, extending even to a planetary scale. Looking through the lens of information theory, the picture of life that emerges is very far from the traditional account of biology, which emphasizes anatomy and physiology.

Many people have helped me in assembling the contents of this book. A lot of the ideas I present here originate with my colleague Sara Walker, Deputy Director of the Beyond Center for Fundamental Concepts in Science, who has greatly influenced my thinking over the past five years. Sara shares my enthusiasm for seeking a grand unified theory of physics and biology organized around the concept of information. 'Life is the next great frontier of physics!' she declares. I have also benefited greatly from discussions with the students and postdocs in our group at Arizona State University (ASU). Special mention must go to Alyssa Adams, Hyunju Kim and Cole Matthis. Among my many brilliant colleagues at ASU, Athena Aktipis, Ariel Anbar, Manfred Laubichler, Stuart Lindsay, Michael Lynch, Carlo Maley, Timothea Newman (now at the University of Dundee) and Ted Pavlic have been especially helpful. Farther afield, I greatly value my many conversations over several years with Christoph Adami at Michigan State University, Gregory Chaitin of the Federal University of Rio de Janeiro, James Crutchfield at the University of California Davis, Andrew Briggs at the University of Oxford, David Chalmers at New York University, Lee Cronin at Glasgow University, Max Tegmark at MIT, Steven Benner at the Foundation for Applied Molecular Evolution, Michael Berry at Bristol University, George Ellis at the University of Cape Town, Piet Hut at the Earth Life Sciences Institute in Tokyo and the Institute for Advanced Study in Princeton, Stuart Kauffman of the Institute of Systems Biology, Charles Lineweaver at the Australian National University, who playfully disagrees with almost everything I say and write, and Christopher McKay at NASA Ames.

Also in Australia, Derek Abbott at the University of Adelaide has clarified several aspects of the physics of life for me, and John Mattick, the visionary director of the Garvan Institute in Sydney, has taught me that genetics and microbiology are not done deals. Paul Griffiths at the University of Sydney has provided me with deep insights into the nature of evolution and epigenetics, while Mikhail Prokopenko and Joe Lizier at the same university have shaped my thinking about network theory and provided some critical feedback. Johnjoe McFadden and Jim Al-Khalili at the University of Surrey, Birgitta Whaley at the University of California, Berkeley, and science writer Philip Ball provided valuable feedback on Chapter 5. Peter Hoffmann of Wayne State University kindly clarified some subtleties about ratchets. Giulio Tononi of the University of Wisconsin, Madison, and his colleagues Larissa Albantakis, and Erik Hoel, now at Columbia University, patiently tried to de-convolve my muddled thinking about integrated information. The Santa Fe Institute has also been a source of inspiration: David Krakauer and David Wolpert have dazzled me with their erudition. Michael Levin at Tufts University is a very valued collaborator and one of the most adventurous biologists I know. I also profited from some lively exchanges with computer engineer and business consultant Perry Marshall.

My foray into cancer research resulted in a rich network of distinguished and brilliant thinkers who have helped shape my understanding of cancer in particular and life in general. At ASU, I have worked closely with Kimberly Bussey and Luis Cisneros on cancer-related projects, and received important help from Mark Vincent at the University of Western Ontario and Robert Austin at Princeton University. My knowledge of cancer genetics was greatly improved by conversations with David Goode and Anna Trigos at the Peter MacCallum Centre in Melbourne, and James Shapiro at the University of Chicago. I have been influenced by the work of Mina Bissell, Brendon Coventry and Thea Tlsty: but there were many others, too numerous to list here. Thanks must also go to the National Cancer Institute, which very generously supported much of the cancer research reported here through a five-year grant, and to NantWorks, which continues to support it. It was the vision of Anna Barker, former deputy director of the National Cancer Institute and now an ASU colleague, that propelled me into cancer research in the first place. In addition, the Templeton World Charity Foundation has strongly supported our origin-of-life research group via their 'Power of Information' programme.

I should also like to thank Penguin Books, my loyal publisher, and particularly Tom Penn, Chloe Currens and Sarah Day for their splendid editorial input.

The final mention must go to Pauline Davies, who carefully read three entire drafts and sent each back for revision, heavily annotated. We have discussed many technical aspects of the book on a daily basis over the past year, and the content is greatly improved by her help. Without her unfailing support for the project, her remorseless cajoling and her razor-sharp intellect, it would never have been satisfactorily completed.

Paul Davies
Sydney and Phoenix, December 2017

I

What *is* Life?

In February 1943 the physicist Erwin Schrödinger delivered a series of lectures at Trinity College in Dublin called *What is Life?* Schrödinger was a celebrity, a Nobel prizewinner, and world famous as an architect of quantum mechanics, the most successful scientific theory ever. It had already explained, within a few years of its formulation in the 1920s, the structure of atoms, the properties of atomic nuclei, radioactivity, the behaviour of subatomic particles, chemical bonds, the thermal and electrical properties of solids and the stability of stars.

Schrödinger's own contribution had begun in 1926 with a new equation that still bears his name, describing how electrons and other subatomic particles move and interact. The decade or so that followed was a golden age for physics, with major advances on almost every front, from the discovery of antimatter and the expanding universe to the prediction of neutrinos and black holes, due in large part to the power of quantum mechanics to explain the atomic and subatomic world. But those heady days came to an abrupt end when in 1939 the world was plunged into war. Many scientists fled Nazi Europe for Britain or the United States to assist the Allied war effort. Schrödinger joined the exodus, leaving his native Austria after the Nazi takeover in 1938, but he decided to make a home in neutral Ireland. Ireland's president, Éamon de Valera, himself a physicist, founded a new Institute for Advanced Studies in 1940 in Dublin. It was de Valera himself who invited Schrödinger to Ireland, where he stayed for sixteen years, accompanied by both wife and mistress living under the same roof.

In the 1940s biology lagged far behind physics. The details of life's basic processes remained largely mysterious. Moreover, the very nature of life seemed to defy one of physics' fundamental laws – the so-called second law of thermodynamics – according to which there is a universal tendency towards degeneration and disorder. In his Dublin lectures Schrödinger set out the problem as he saw it: 'how can the events in space and time which take place within the spatial boundary of a living organism be accounted for by physics and chemistry?' In other words, can the baffling properties of living organisms ultimately be reduced to atomic physics, or is something else going on? Schrödinger put his finger on the key issue. For life to generate order out of disorder and buck the second law of thermodynamics, there had to be a molecular entity that somehow *encoded* the instructions for building an organism, at once complex enough to embed a vast quantity of information and stable enough to withstand the degrading effects of thermodynamics. We now know that this entity is DNA.

In the wake of Schrödinger's penetrating insights, which were published in book form the next year, the field of molecular biology exploded. The elucidation of the structure of DNA, the cracking of the genetic code and the merging of genetics with the theory of evolution followed swiftly. So rapid and so sweeping were the successes of molecular biology that most scientists adopted a strongly reductionist view: it did indeed seem that the astonishing properties of living matter could ultimately be explained solely in terms of the physics of atoms and molecules, without the need for anything fundamentally new. Schrödinger himself, however, was less sanguine: '. . . living matter, while not eluding the "laws of physics" as established up to date, is likely to involve "other laws of physics" hitherto unknown . . .' he wrote.[1] In this he was not alone. Fellow architects of quantum mechanics such as Niels Bohr and Werner Heisenberg also felt that living matter might require new physics.

Strong reductionism still prevails in biology. The orthodox view remains that known physics alone is all that is needed to explain life, even if most of the details haven't been entirely worked out. I disagree. Like Schrödinger, I think living organisms manifest deep new physical principles, and that we are on the threshold of uncovering and harnessing those principles. What is different this time, and why it has taken so many decades to discover the real secret of life, is that the new physics is not simply a matter of an additional type of force – a 'life force' – but something altogether more subtle, something that interweaves matter and information, wholes and parts, simplicity and complexity.

That 'something' is the central theme of this book.

Box 1: The magic puzzle box

Ask the question 'What is life?' and many qualities clamour for our attention. Living organisms reproduce themselves, explore boundless novelty through evolution, invent totally new systems and structures by navigating possibility space along trajectories impossible to predict, use sophisticated algorithms to compute survival strategies, create order out of chaos, going against the cosmic tide of degeneration and decay, manifest clear goals and harness diverse sources of energy to attain them, form webs of unimaginable complexity, cooperate and compete ... the list goes on. To answer Schrödinger's question we have to embrace *all* these properties, connecting the dots from across the entire scientific spectrum into an organized theory. It is an intellectual adventure that interweaves the foundations of logic and mathematics, paradoxes of self-reference, the theory of computation, the science of heat engines, the burgeoning achievements of nanotechnology, the emerging field of far-from-equilibrium thermodynamics and the enigmatic domain of quantum physics. The unifying feature of all these subjects is *information*, a concept at once familiar and pragmatic but also abstract and mathematical, lying at the foundation of both biology and physics.

Charles Darwin famously wrote, 'It is interesting to contemplate a tangled bank, clothed with many plants of many kinds, with birds singing on the bushes, with various insects flitting about, and with worms crawling through the damp earth, and to reflect that these elaborately constructed forms, so different from each other, and dependent upon each other in so complex a manner, have all been produced by laws acting around us.'[2] What Darwin didn't envisage is that threading through this manifest material complexity (the hardware of life) was an even more breathtaking *informational* complexity (the software of life), hidden from view but providing the guiding hand of both adaptation and novelty. It is here in the realm of information that we encounter life's true creative power. Now, scientists are merging the hardware and software narratives into a new theory of life that has sweeping ramifications from astrobiology to medicine.

GOODBYE LIFE FORCE

Throughout history it was recognized that living organisms possess strange powers, such as the ability to move autonomously, to rearrange their environment and to reproduce. The philosopher Aristotle attempted to capture this elusive otherness with a concept known as *teleology* – derived from the Greek word *telos*, meaning 'goal' or 'end'. Aristotle observed that organisms seem to behave purposefully according to some pre-arranged plan or project, their activities being directed or pulled towards a final state, whether it is seizing food, building a nest or procreating through sex.

In the early scientific era the view persisted that living things were made of a type of magic matter, or at least normal matter infused with an added ingredient. It was a point of view known as *vitalism*. Just what that extra essence might be was left vague; suggestions included air (the breath of life), heat, electricity, or something mystical like the soul. Whatever it might be, the assumption that a special type of 'life force' or ethereal energy served to animate matter was widespread into the nineteenth century.

With improvements in scientific techniques such as the use of powerful microscopes, biologists found more and more surprises that seemed to demand a life force. One major puzzle concerned embryo development. Who could not be astonished by the way that a single fertilized egg cell, too small to see with the unaided eye, can grow into a baby? What guides the embryo's complex organization? How can it unfold so reliably to produce such an exquisitely arranged outcome? The German embryologist Hans Dreisch was particularly struck by a series of experiments he performed in 1885. Dreisch tried mutilating the embryos of sea urchins – a favourite victim among biologists – only to find they somehow recovered and developed normally. He discovered it was even possible to disaggregate the developing ball of cells at the four-cell stage and grow each individual cell into a complete sea urchin. Results such as these gave Dreisch the impression that the embryonic cells possessed some 'idea in advance' of the final shape they intended to create and cleverly compensated for the experimenter's meddling. It was as if some invisible hand supervised their growth and development, effecting 'mid-course corrections' if necessary. To Dreisch, these facts constituted strong evidence for some form of vital essence, which he termed *entelechy*, meaning 'complete, perfect, final form' in Greek, an idea closely related to Aristotle's notion of teleology.

But trouble was brewing for the life force. For such a force actually to accomplish something it must – like all forces – be able to move matter. And at first sight, organisms do indeed seem to be self-propelled, to possess some inner source of motive power. But exerting any kind of force involves expending energy. So, if the 'life force' is real, then the transfer of energy should be measurable. The physicist Hermann von Helmholtz investigated this very issue intensively in the 1840s. In a series of experiments he applied pulses of electricity to muscles extracted from frogs, which caused them to twitch, and carefully measured the minute changes in temperature that accompanied the movement. Helmholtz concluded that it was chemical energy stored in the muscles that, triggered by the jolt of electricity, became converted into the mechanical energy of twitching which, in turn, degraded into heat. The energy books balanced nicely without any evidence of the need for additional vital forces to be deployed. Yet it took several more decades for vitalism to fade away completely.*

But even without a life force, it's hard to shake the impression that there is *something* special about living matter. The question is, what?

I became fascinated with this conundrum after reading Schrödinger's book *What is Life?* as a student. At one level the answer is straightforward: living organisms reproduce, metabolize, respond to

---

* Vitalism also suffered from being too close in concept to another nineteenth-century fad – spiritualism, with its bizarre stories of ectoplasm and aetheric bodies.

stimuli, and so forth. However, merely listing the properties of life does not amount to an *explanation*, which is what Schrödinger sought. As much as I was inspired by Schrödinger's book, I found his account frustratingly incomplete. It was clear to me that life must involve more than just the physics of atoms and molecules. Although Schrödinger suggested that some sort of new physics might be at play, he didn't say what. Subsequent advances in molecular biology and biophysics gave few clues. But very recently the outline of a solution has emerged, and it comes from a totally novel direction.

LIFE SPRINGS SURPRISES

'Base metals can be transmuted into gold by stars, and by intelligent beings who understand the processes that power stars, and by nothing else in the universe.'

– *David Deutsch*[3]

Understanding the answer to Schrödinger's question 'What is life?' means abandoning the traditional list of properties that biologists reel off, and beginning to think about the living state in a totally new way. Ask the question, 'How would the world be different if there were no life?' It is common knowledge that our planet has been shaped in part by biology: the build-up of oxygen in the atmosphere, the formation of mineral deposits, the worldwide effects of human technology. Many non-living processes also reshape the planet – volcanic eruptions, asteroid impacts, glaciation. The key distinction is that life brings about processes that are not only unlikely but *impossible* in any other way. What else can fly halfway round the world with pinpoint precision (the Arctic tern), convert sunlight into electrical energy with 90 per cent efficiency (leaves) or build complex networks of underground tunnels (termites)?

Of course, human technology – a product of life – can do these things too, and more. To illustrate: for 4.5 billion years, since the solar system formed, Earth has accumulated material – the technical term is 'accretion' – from asteroid and comet impacts. Objects of various sizes, from hundreds of kilometres across to tiny meteoric grains, have rained down throughout our planet's history. Most people know about the dinosaur-destroying comet that slammed into what is now Mexico 65 million years ago, but that was just one instance. Aeons of bombardment means that our planet is slightly heavier today than it was in the past. Since 1958, however, 'anti-accretion' has occurred.[4] Without any sort of geological catastrophe, a large swarm of objects has gone the other way – up from Earth into space, some to travel to the moon and planets, others to journey into the void for good; most of them have ended up orbiting Earth. This state of affairs would be impossible based purely on the laws of mechanics and planetary evolution. It is readily explained, however, by human rocket technology.

Another example. When the solar system formed, a small fraction of its initial chemical inventory included the element plutonium. Because the longest-lived isotope of plutonium has a half-life of about 81 million years, virtually all the primordial plutonium has now decayed. But in 1940 plutonium reappeared on Earth as a result of experiments in nuclear physics; there are now estimated to be a thousand tonnes of it. Without life, the sudden rise of terrestrial plutonium would be utterly inexplicable. There is no plausible non-living pathway from a 4.5-billion-year-old dead planet to one with deposits of plutonium.

Life does not merely effect these changes opportunistically, it diversifies and adapts, invading new niches and inventing ingenious mechanisms to make a living, sometimes in extraordinary ways. Three kilometres below ground in South Africa's Mponeng gold mine colonies of exotic bacteria nestle in the microscopic pores of the torrid, gold-bearing rocks, isolated from the rest of our planet's biosphere. There is no light to sustain them, no organic raw material to eat. The source of the microbes' precarious existence is, astonishingly, radioactivity. Normally deadly to life, nuclear radiation emanating from the rocks provides the subterranean denizens with enough energy by splitting water into oxygen and hydrogen. The bacteria, known as *Desulforudis audaxviator*, have evolved mechanisms to exploit the chemical by-products of radiation, making biomass by combining the hydrogen with carbon dioxide dissolved in the scalding water that suffuses the rocks.

Eight thousand kilometres away, in the desiccated heart of the Atacama Desert in Chile, the fierce sun rises over a unique landscape. As far as the eye can see there is only sand and rock, unrelieved by signs of life. No birds, insects or plants embellish the view. Nothing scrambles in the dust, no green patches betray the presence of even simple algae. All known life needs liquid water, and it virtually never rains in this region of the Atacama, making it the driest, and deadest, place on the Earth's surface.

The core of the Atacama is Earth's closest analogue to the surface of Mars, so NASA has a field station there to test theories about Martian soil. The scientists went originally to study the outer limits of life – they like to say they are looking for death, not life – but what they found instead was startling. Scattered amid the outcrops of desert rock are weird, sand-encrusted shapes, pillars rising to a height of a metre or so, rounded and knobbed, resembling a riot of sculptures that might have been designed by Salvador Dalí. The mounds are in fact made of salt, remnants of an ancient lake long since evaporated. And inside the pillars, literally entombed in salt, are living microbes that eke out a desperate existence against all the odds. These very different weird organisms, named *Chroococcidiopsis*, get their energy not from radioactivity but, more conventionally, from photosynthesis; the strong desert sunlight penetrates their translucent dwellings. But there remains the question of water. This part of the Atacama Desert lies about a hundred kilometres inland from the cold Pacific Ocean, from which it is separated by a mountain range. Under the right conditions, fingers of sea mist meander through the mountain passes at night when the temperature plunges. The dank air infuses water molecules into the salt matrix. The water doesn't form liquid droplets; rather, the salt becomes damp and sticky, a phenomenon well known to those readers who live in wet climes and are familiar with obstinate salt cellars. The absorption of water vapour into salt is called deliquescence, and it serves well enough – just – to keep the microbes happy for a while before the morning sun bakes the salt dry.

Desulforudis audaxviator and *Chroococcidiopsis* are two examples illustrating the extraordinary ability of living organisms to survive in dire circumstances. Other microbes are known to withstand extremes of cold or heat, salinity and metal contamination, and acidity fierce enough to burn human flesh. The discovery of this menagerie of resilient microbes living on the edge (collectively called extremophiles) overturned a long-standing belief that life could flourish only within narrow margins of temperature, pressure, acidity, and so forth. But life's profound ability to create new physical and chemical pathways and to tap into a range of unlikely energy sources illustrates how, once life gets going, it has the potential to spread far beyond its original habitat and trigger unexpected transformations. In the far future, humans or their machine descendants may reconfigure the entire solar system or even the galaxy. Other forms of life elsewhere in the universe could already be doing something similar, or may eventually do so. Now that life has been unleashed into the universe, it has within it the potential to bring about changes of literally cosmic significance.

THE VEXING PROBLEM
OF THE LIFE METER

There is a dictum in science that if something is real, then we ought to be able to measure it (and perhaps even tax it). Can we measure life? Or 'degree of aliveness'? That may seem an abstract question, but it recently assumed a certain immediacy. In 1997 the US and European space agencies collaborated to send a spacecraft named *Cassini* to Saturn and its moons. Great interest was focused on the moon Titan, the largest in the solar system. Titan was discovered by Christiaan Huygens in 1655 and has long been a curiosity to astronomers, not only because of its size but because it is covered in clouds. Until the *Cassini* mission, what lay beneath was, literally, shrouded in mystery. The *Cassini* spacecraft conveyed a small probe, fittingly called Huygens, which was dropped through Titan's clouds to land safely on the moon's surface. Huygens revealed a landscape featuring oceans and beaches, but the oceans are made of liquid ethane and methane and the rocks are made of water ice. Titan is very cold, with an average temperature of $-180°C$.

Astrobiologists took a keen interest in the *Cassini* mission. It was already known that Titan's atmosphere is a thick petrochemical smog, its clouds replete with organic molecules. However, due to the extreme cold, this body could not sustain life as we know it. There has been some speculation that an exotic form of life might be able to use liquid methane as a substitute for water, though most astrobiologists do not think that likely. However, even if Titan is completely dead, it is still very relevant to the puzzle of life. In effect, it constitutes a natural chemistry laboratory that has been steadily cooking up complex organic molecules for its entire lifespan of 4.5 billion years. Put more colourfully, Titan is a kind of gigantic failed biology experiment – which gets us right to the heart of the 'what is life?' problem. If Titan has travelled, chemically speaking, part-way down the long and convoluted path that on Earth culminated in life, *how close* has Titan come to the finishing line labelled 'biology begins here'? Could it be that Titan is now in some sense fairly close to incubating life? Is there such a thing as 'almost life' that we might discover lurking in its murky clouds?

Put more starkly, is it possible to build some sort of life meter that can sample Titan's organic-laden atmosphere and deliver a number? Imagine a future mission followed by an announcement from the scientific team: 'Over a period of 4.5 billion years the smog on Titan succeeded in getting 87.3 per cent of the way to life.' Or perhaps, 'Titan managed to travel only 4 per cent of the long journey from organic building blocks to a simple living cell.'

These statements sound ridiculous. But why?

Of course, we do not possess a life meter. More to the point, it is very unclear how such a device would even work in principle. What exactly would it measure? Richard Dawkins introduced an engaging metaphor to illustrate the process of biological evolution, called Mount Improbable.[5] Complex life is a priori exceedingly unlikely. It exists only because, starting with very simple microbial organisms, evolution by natural selection has incrementally fashioned it over immense periods of time. In the metaphor, the ancestors of today's complex life forms (such as humans) can be envisaged as climbing higher and higher up the mountain (in the complexity sense) over billions of years. Fair enough. But what about the first step, the transformation from non-life to life, the road from a mishmash of simple chemicals to a primitive living cell? Was that also a climb up a sort of prebiotic, chemical version of Mount Improbable? It seems it must be so. The transition from a random mixture of simple molecules to a fully functional organism obviously didn't happen in one huge, amazing chemical leap. There must have been a long journey through intermediate steps. Nobody knows what those steps were (apart perhaps from the very first ones; see p. 167). In fact, we don't know the answer to an even more basic question: was the ascent from non-life to life a long, gently sloping, seamless upward track from inanimate matter to biology, or did it feature a series of abrupt major transformations, akin to what in physics are called phase transitions (such as the jump from water to steam)? Nobody knows. In either case, however, the metaphor of a prebiotic Mount Improbable is useful, with the height up the mountain a measure of chemical complexity. Returning to the hypothetical life meter to be sent to Titan, if it existed, it could be regarded as a sort of complexity altimeter that would measure how far up the prebiotic Mount Improbable Titan's atmosphere had climbed.

Clearly something is missing in an account that focuses on chemical complexity alone. A recently deceased mouse is chemically as complex as a living mouse, but we wouldn't think of it as being, say, 99.9 per cent alive. It is simply dead.* And what about the case of dormant but not actually dead microbes, such as bacteria that form spores when confronting adverse conditions, remaining inert until they encounter better circumstances and 'start ticking' again? Or the tiny eight-legged animals called tardigrades (water bears), which when cooled to liquid-helium temperatures have been found to simply shut down yet revert to business as usual when they are warmed up again? There will of course be limits to the viability of even these resilient organisms. Could a life meter tell us when a bacterial spore or a tardigrade has gone beyond the point of no return and will 'never wake up'?

The issue is not merely a philosophical poser. Saturn has another icy moon that has received a lot of attention in recent years. Called Enceladus, it is heated from within by tidal flexing of its solid core, brought about as the moon orbits the giant planet. So, although Enceladus is very far from the sun and has a frozen surface, beneath its icy crust lies a liquid ocean. The crust is not, however, perfectly intact. *Cassini* found that Enceladus is spewing material into space from gigantic fissures in the ice. And among the substances

---

* This is a simplification: different organs die at different rates, and the bacteria that inhabit the mouse may live on for ever.

emanating from the interior are organic molecules. Do they hint at life lurking beneath the frigid surface? How could we tell?

NASA is planning a mission to fly through a plume of Enceladus in the 2020s with the express purpose of looking for traces of biological activity. But there is a pressing question: what instruments should go on this probe and what should they look for? Can we design a life meter for the journey? Even if it isn't possible to measure the 'degree of aliveness' precisely, could an instrument at least tell the difference between 'far from life', 'almost alive', 'alive', and 'once living but now dead'? Is that question even meaningful in the form stated?

The life-meter difficulty points at a wider problem. There is great excitement about the prospect of studying the atmospheres of extra-solar planets in enough detail to reveal the telltale signs of life at work. But what would be a convincing smoking gun? Some astro-biologists think atmospheric oxygen would be a giveaway, implying photosynthesis; others suggest methane, or a mixture of the two gases. In truth, there is no agreement, because all the common gases can be produced by non-biological mechanisms too.

A salutary lesson in the perils of defining life in advance came in 1976 when two NASA spacecraft called *Viking* landed on Mars. It was the first and last time that the US space agency attempted to do actual biological experiments on another planet, as opposed to simply studying whether the *conditions* for life may exist. One of the *Viking* experiments, Labelled Release, was designed by engineer Gil Levin, now an Adjunct Professor at ASU. It worked by pouring a nutrient medium onto some Mars dirt to see whether the broth was consumed by any resident microbes and converted into carbon dioxide waste. The carbon in the broth was radioactively tagged so it could be spotted if it emerged as CO_2. And it was indeed spotted. Furthermore, when the sample was baked, the reaction stopped, as it would if Martian microbes had been killed by the heat. The same results were obtained on both *Viking* craft at different locations on Mars, and in repeated runs of the experiment. To this day, Gil claims he detected life on Mars and that history will eventually prove him right. NASA's official pronouncement, by contrast, is that *Viking* did *not* find life, and that the Labelled Release results were due to unusual soil conditions. Probably that's why the agency has never felt motivated to repeat the experiment.

This sharp disagreement between NASA and one of its mission scientists shows how hard it is in practice to decide whether life is present on another world if we have only chemistry to go on. *Viking* was designed to look for chemical traces of life as we know it. If we could be certain that terrestrial life was the only sort possible, we could design equipment to detect organic molecules sufficiently complex that they could be produced only by known biology. If the equipment found, say, a ribosome (a molecular machine needed to make proteins), biologists would be convinced that the sample was either alive now, or had been alive in the near past. But how about simpler molecules used by known life such as amino acids? Not good enough: some meteorites contain amino acids that formed in space without the need for biological processes. Recently, the sugar glycolaldehyde was discovered in a gas cloud near a star 400 light years away, but on its own it is a far cry from a clear signature of life, as such molecules can form from simple chemistry. So it's possible to bracket the *range* of chemical complexity, but where along the line of molecules from amino acids and sugars to ribosomes and proteins could one say that life was *definitely* involved? Is it even possible to identify life *purely from its chemical fingerprint*?* Many scientists prefer to think of life as a *process* rather than a thing, perhaps as a process that makes sense only on a planetary scale.[6] (See Box 12 in Chapter 6.)

THE TALE OF THE ANCIENT MOLECULE

Some type of life has existed on Earth for about 4 billion years. During that time there have been asteroid and comet bombardments, massive volcanism, global glaciation and an inexorably warming sun. Yet life in one form or another has flourished. The common thread running through the story of life on Earth – in this case, literally – is a long

---

* Lee Cronin of the University of Glasgow has proposed a measure of chemical complexity based on the number of steps needed to produce a given large molecule.

molecule called DNA, discovered in 1869 by the Swiss chemist Friedrich Miescher. Derived from the Latin word *moles*, the term 'molecule'
came into vogue in eighteenth-century France to mean 'extremely
small mass'. Yet DNA is anything but small. Every cell in your body
contains about two metres of it – it's a giant among molecules. Etched
into its famous double-helix structure is the instruction manual for life.
The basic recipe is the same for all known life; we share 98 per cent of
our genes with chimpanzees, 85 per cent with mice, 60 per cent with
chickens and more than half with many bacteria.

Box 2: Life's basic machinery

The informational basis of all life on Earth is the universal genetic
code. The information needed to build a given protein is stored in segments of DNA as a specific sequence of 'letters': A, C, G, T. The letters
stand for the molecules adenosine, cytosine, guanine and thymine,
collectively known as bases, and they can be arranged in any combination along the DNA molecule. Different combinations code for
different proteins. Proteins are made from other types of molecules
called amino acids; a typical protein will consist of hundreds of amino
acids linked end to end to form a chain. There are many amino acids,
but life as we know it uses only a restricted set of twenty (sometimes
twenty-one). The chemical properties of a protein will depend on the
precise sequence of amino acids. Because there are only four bases but
twenty amino acids, DNA cannot use a single base to specify each
amino acid. Instead, it uses groups of three in a row. There are sixty-
four possible triplet combinations, or codons, of the four letters (for
example, ACT, GCA . . .). Sixty-four is more than enough for twenty
amino acids, so there is some redundancy: many amino acids are specified by two or more different codons. A few codons are used for
punctuation (for example, 'stop').

To 'read out' the instructions for building a given protein, the cell
first transcribes the relevant codon sequence from DNA into a related
molecule called mRNA (messenger RNA). Proteins are assembled by

ribosomes, little machines that read off the sequence of codons from mRNA and synthesize the protein step by step by chemically linking together one amino acid to another. Each codon has to get the right amino acid for the system to work properly. That is achieved with the help of another form of RNA (transfer RNA, or tRNA for short). These short strands of RNA come in twenty varieties, each customized to recognize a specific codon and bind to it. Crucially, hanging on to this tRNA is the appropriate amino acid to match the codon that codes for it, waiting to be delivered to the growing chain of amino acids that will make up the functional protein when the ribosome has finished. For all this to work, the right amino acid from the set of twenty has to be attached to the corresponding variety of tRNA. This step happens care of a special protein with the daunting name of aminoacyl-tRNA synthetase. The name doesn't matter. What does matter is that the shape of this protein is specific to *both* tRNA and to the corresponding amino acid so it can attach the correct amino acid onto the corresponding variety of tRNA. As there are twenty different amino acids, there have to be twenty different aminoacyl-tRNA synthetases. Notice that aminoacyl-tRNA synthetases are the crucial link in the information chain. Biological information is stored in one type of molecule (DNA, a nucleic acid, which uses a triplet code with a four-letter alphabet), but it is intended for a completely different class of molecule (proteins, which use a twenty-letter alphabet). These two types of molecule speak a different language! But the set of aminoacyl-tRNA synthetases are bilingual: they recognize both codons and the twenty varieties of amino acids. These linking molecules are therefore absolutely critical to the universal genetic machinery that all known life uses. Consequently, they must be very ancient and had better work very well. All life depends on it! Experiments show that they are indeed extremely reliable, getting it wrong (that is, bungling the translation) in only about 1 in 3,000 cases. It is hard not to be struck by how *ingenious* all this machinery is, and how astonishing that it remains intact and unchanged over billions of years.

The fact that all known life follows a universal script suggests a common origin. The oldest traces of life on Earth date back at least 3.5 billion years, and it is thought that some portions of DNA have remained largely unchanged in all that time. Also unchanged is the language of life. The DNA rule book is written in code, using the four letters A, C, G and T for the four nucleic acid bases which, strung together on a scaffold, make up the structure of the ancient molecule.* The sequence of bases, when decoded, specifies the recipes for building proteins: the workhorses of biology. Human DNA codes for about 20,000 of them. Although organisms may have different proteins, all share the same coding and decoding scheme (Box 2 has details). Proteins are made from strings of amino acids hooked together. A typical protein consists of a chain of several hundred amino acids folded into a complex three-dimensional shape – its functional form. Life uses twenty (sometimes twenty-one) varieties of amino acids in various combinations. There are countless ways that sequences of A, C, G and T bases can code for twenty amino acids, but all known life uses the *same* assignments (see Table 1), suggesting it is a very ancient and deeply embedded feature of life on Earth, present in a common ancestor billions of years ago.†

Although DNA is very ancient, other entities have staying power too: crystals, for example. There are zircons in Australia and Canada that have been around for over 4 billion years and have survived episodes of subduction into the Earth's crust. The chief difference is that a living organism is out of equilibrium with its environment. In fact, life is generally very *far* out of equilibrium. To continue to function, an organism has to acquire energy from the environment (for example, from sunlight or by eating food) and export something (for example, oxygen or carbon dioxide). There is thus a continuous exchange of energy and material with the surroundings, whereas a crystal is internally inert. When an organism dies, all that activity stops, and it steadily slides into equilibrium as it decays.

---

* When information from DNA is transcribed into RNA, T is substituted for a slightly different molecule labelled U, which stands for uracil.
† A terminological point: scientists cause much confusion when they refer to an organism's 'code' when they really mean coded genetic data. Your genetic data and mine differ, but we have the same genetic code.

Table 1: The universal genetic code

	T		C		A		G	
T	TTT	Phe	TCT	Ser	TAT	Tyr	TGT	Cys
	TTC		TCC		TAC		TGC	
	TTA	Leu	TCA		TAA	STOP	TGA	STOP
	TTG		TCG		TAG		TGG	Trp
C	CTT	Leu	CCT	Pro	CAT	His	CGT	Arg
	CTC		CCC		CAC		CGC	
	CTA		CCA		CCA	Gln	CGA	
	CTG		CCG		CAG		CGG	
A	ATT	Ile	ACT	Thr	AAT	Asn	AGT	Ser
	ATC		ACC		AAC		AGC	
	ATA		ACA		AAA	Lys	AGA	Arg
	ATG	Met	ACG		AAG		AGG	
G	GTT	Val	GCT	Ala	GAT	Asp	GGT	Gly
	GTC		GCC		GAC		GGC	
	GTA		GCA		GAA	Glu	GGA	
	GTG		GCG		GAG		GGG	

The table shows the coding assignments used by all known life. The amino acids which the triplets of letters (codons) code for are listed to the right of the codons, as abbreviations (e.g. Phe = phenylalanine; the names of all these molecules are unimportant for my purposes). A historical curiosity: the existence of some form of genetic code was originally suggested in a letter to Crick and Watson dated 8 July 1953 by a cosmologist, George Gamow, better known for his pioneering work on the Big Bang.

There are certainly non-living systems that are also far out of equilibrium and likewise have good staying power. My favourite example is the Great Red Spot of Jupiter, which is a gaseous vortex that has endured ever since the planet has been observed through telescopes and shows no sign of going away (see Fig. 1). Many other examples are known of chemical or physical systems with a similar type of

Fig. 1. Jupiter's Great Red Spot.

autonomous existence. One of these is convection cells, in which a fluid (for example, liquid water) rises and falls in a systematic pattern when heated from below. Then there are chemical reactions that generate spiral shapes or pulsate rhythmically (Fig. 2). Systems like these which display the spontaneous appearance of organized complexity were dubbed 'dissipative structures' by the chemist Ilya Prigogine, who championed their study in the 1970s. Prigogine felt that chemical dissipative structures, operating far from equilibrium and supporting a continued throughput of matter and energy, represented a sort of waystation on the long road to life. Many scientists believe that still.

In living things, most chemical activity is handled by proteins.

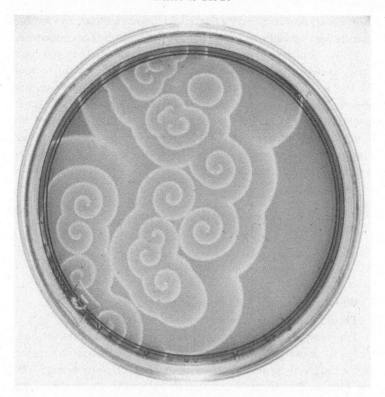

Fig. 2. A chemical 'dissipative structure'. When a particular chemical mixture is forced away from equilibrium it can spontaneously evolve stable forms of the type shown. The chemist Ilya Prigogine maintained that such systems represent the first steps on a long road to life.

Metabolism – the flow of energy and material through organisms – is necessary for life to achieve anything, and proteins do the lion's share of metabolic work. If life got started (as Prigogine believed) via elaborate energy-driven chemical cycles, then proteins must have been early actors in the great drama of life. But on their own, proteins are largely useless. The all-important *organization* of life requires a great deal of choreography, that is, some form of command-and-control arrangement. That job is done by nucleic acids (DNA and RNA).

Life as we know it involves a deal struck between these two very different classes of molecules: nucleic acids and proteins. The conundrum as most scientists see it is the chicken-and-egg nature of life: you can't have one without the other. Without a legion of proteins to fuss around it, a molecule of DNA is stranded. In simplistic terms, the job descriptions are: nucleic acids store the details about the 'life plan'; proteins do the grunt work to run the organism. *Both* are needed. So a definition of life must take this into account. It needs to consider not just complex pattern-creating organized chemistry but *supervised* or *informed* chemistry: in short, chemistry plus *information*.

LIFE = MATTER + INFORMATION

'Nothing in biology makes sense except in the light of information.'

Bernd-Olaf Küppers[7]

We have now arrived at a critical juncture.

The thing that separates life from non-life is *information*.

That's easy to state, but it needs some unpacking. Start with something simple: organisms reproduce and in so doing they pass on information about their form to their offspring. In that respect, reproduction is not the same as production. When dogs reproduce they make more dogs; cats make cats, humans make humans. The basic body plan propagates from one generation to the next. But reproduction is more nuanced than mere species perpetuation. Human babies, for example, inherit some detailed characteristics from their parents or grandparents – red hair, blue eyes, freckles, long legs . . . The best way to express inheritance is to say that *information* about earlier generations is passed along to the next – the information needed to build a new organism in the likeness of the old. This information is encoded in the organism's genes, which are replicated as part of the reproductive process. The essence of biological reproduction, then, is the replication of *heritable information*.

When Schrödinger gave his lectures in 1943 scientists were mostly in the dark about how genetic information was copied and passed on. Nobody really knew where this information was stored or how it was replicated. This was a decade before the discovery of the role of DNA in genetics. Schrödinger's great insight was to identify how information storage, processing and transmission must take place at the *molecular* level, on a nanoscale, within living cells.* Furthermore, quantum mechanics – Schrödinger's own brainchild – was needed to explain the stability of the information storage. Although the genetic material was unknown, Schrödinger concluded it would involve a molecule with a definite structure which he termed 'an aperiodic crystal'. It was an extremely perceptive suggestion. A crystal has stability. But familiar crystals, like diamonds or salt, are periodic: simple, regular arrays of atoms. On the other hand, a molecule with crystalline levels of stability that could be *arbitrarily* structured might encode and store a vast amount of information. And that is precisely what DNA turned out to be: an aperiodic crystal. Both Crick and Watson, who discovered the structure of DNA a decade later, acknowledged that Schrödinger's book had given them essential food for thought in elucidating the form and function of the elusive genetic material.

Today, the informational basis of life has permeated every aspect of science. Biologists say that genes (definite sequences of bases in DNA) contain 'coded instructions' that are 'transcribed' and 'translated'. When genes are replicated, information is first copied and then proof-read; errors are corrected if necessary. On the scale of tissues, 'signalling' molecules communicate information between neighbouring cells; other molecules circulate in the blood, sending signals between organs. Even single cells gather information about their environment, process it internally and respond accordingly. The pre-eminent information-processing system in biology is of course the brain, often compared (not very convincingly) to a digital computer. And beyond individual organisms lie social structures and ecosystems. Social insects like ants and bees transfer information to help them coordinate group activities such as foraging and nest-site selection. Birds aggregate into flocks and fish into shoals: information exchange lies at the heart of their coordinated behaviour. Primates organize themselves into colonies with complex social norms maintained by many subtle forms of communication. Human society

---

* A nanometre is one billionth of a metre. 'Nanotechnology' refers to engineered structures on this molecular scale.

has spawned planet-wide information-processing systems like the World Wide Web. It is thus no surprise that many scientists now choose to *define* life in terms of its informational properties: 'a chemical system in which the flow and storage of energy are related to the flow and storage of information' is the way biophysicist Eric Smith expresses it.[8]

We have now reached the nexus at which the disparate realms of biology and physics, of life and non-life, meet. Although Schrödinger correctly put his finger on the existence of a link between molecular structure and information storage, his aperiodic crystal proposal glossed over a vast conceptual chasm. A molecule is a physical structure; information, on the other hand, is an abstract concept, deriving ultimately from the world of human communication. How can the chasm be bridged? How can we link abstract information to the physics of molecules? The first glimmerings of an answer came, as it happened, 150 years ago in the ferment of the Industrial Revolution, and it came from a subject that had less to do with biology and more with the nuts-and-bolts field of mechanical engineering.

2

Enter the Demon

'Could life's machines be Maxwell demons, creating order out of chaos . . . ?'

– *Peter Hoffmann*[1]

In December 1867 the Scottish physicist James Clerk Maxwell penned a letter to his friend Peter Guthrie Tait. Though little more than speculative musing, Maxwell's missive contained a bombshell that still reverberates a century and a half later. The source of the disruption was an imaginary being – 'a being whose faculties are so sharpened that he can follow every molecule in its course'. Using a simple argument, Maxwell concluded that this Lilliputian entity, soon to be dubbed a *demon*, 'would be able to do what is impossible to us'. On the face of it, the demon could perform magic, conjuring order out of chaos and offering the first hint of a link between the abstract world of information and the physical world of molecules.

Maxwell, it should be stressed, was an intellectual giant, in stature comparable to Newton and Einstein. In the 1850s he unified the laws of electromagnetism and predicted the existence of radio waves. He was also a pioneer of colour photography and explained Saturn's rings. More relevantly, he made seminal contributions to the theory of heat, calculating how, in a gas at a given temperature, the heat energy was shared out among the countless chaotically moving molecules.

Maxwell's demon was a paradox, an enigma, an affront to the lawfulness of the universe. It opened a Pandora's box of puzzles about the nature of order and chaos, growth and decay, meaning and purpose. And although Maxwell was a physicist, it turned out that the most powerful application of the demon idea lay not in physics but in biology. Maxwell's demonic magic can, we now know, help explain the magic of life. But that application lay far in the future. At the outset, the demon wasn't intended to clarify the question 'What is life?' but a much simpler and more practical one: namely, what is heat?

MOLECULAR MAGIC

Maxwell wrote to Tait at the height of the Industrial Revolution. Unlike the agricultural revolution of the Neolithic period which pre-dated it by several thousand years, the Industrial Revolution did not proceed by trial and error. Machines such as the steam engine and the diesel engine were carefully designed by scientists and engineers familiar with the principles of mechanics first enunciated by Isaac Newton in the seventeenth century. Newton had discovered the laws of motion, which relate the force acting on a material body to the nature of its movement, all encapsulated in a simple mathematical formula. By the nineteenth century it was commonplace to use Newton's laws to design tunnels and bridges or to predict the behaviour of pistons and wheels, the traction they would deliver and the energy they would need.

By the middle of the nineteenth century physics was a mature science, and the welter of engineering problems thrown up by the new industries provided fascinating challenges for physicists to analyse. The key to industrial growth lay, then as now, with energy. Coal provided the readiest source to power heavy machinery, and steam engines were the preferred means of turning the chemical energy of coal into mechanical traction. Optimizing the trade-off between energy, heat, work and waste was more than just an academic exercise. Vast profits could hinge on a modest improvement in efficiency.

Although the laws of mechanics were well understood at the time, the nature of heat remained confusing. Engineers knew it was a type of energy that could be converted into other forms, for example, into the energy of motion – the principle behind the steam locomotive. But harnessing heat to perform useful work turned out to involve more than a simple transfer between different forms of energy. If we had unrestricted access to all heat energy, the world would be a very different place, because heat is a highly abundant source of energy in the universe.* The unrestricted exploitation of heat energy would, for instance, enable a spacecraft to be propelled entirely from the thermal afterglow of the Big Bang. Or, coming closer to home, we could power all our industries on water alone: there is enough heat energy in a bottle of water to illuminate my living room for an hour. Imagine sailing a ship with no fuel other than the heat of the ocean.

Sadly, it can't be done. Pesky physicists discovered in the 1860s a strict limit on the amount of heat that can be converted into useful mechanical activity. The constraint stems from the fact that it is the *flow* of heat, not heat energy per se, that can perform work. To harness heat energy there has to be a temperature *difference* somewhere. Simple example: if a tank of hot water is placed near a tank of cold water, then a heat engine connected to both can run off the temperature gradient and perform a physical task like turning a flywheel or lifting a weight. The engine will take heat from the hot water and deliver it to the cold water, extracting some useful energy on the way. But as heat is transferred from the hot tank to the cold tank, the hot water will get cooler and the cold water will get warmer, until the temperature difference between the two dwindles and the motor grinds to a halt. What is the best-case scenario? The answer depends on the temperatures of the tanks, but if one tank is maintained (by some external equipment) at boiling point (100°C) and the other at freezing point (0°C), then it turns out that the best one can hope for – even if no heat is wasted by leaking into the surroundings – is to extract about 27 per cent of the heat energy in the form of useful work. No engineer in the universe could better that; it is a fundamental law of nature.

Once physicists had figured this out, the science known as thermodynamics was born. The law that says you can't convert all the heat energy into work is the second law of thermodynamics.† This same

---

* I'm ignoring here the mass-energy of matter, which is mostly inert, and the mysterious dark energy of empty space. They are far more abundant.
† The first law of thermodynamics is just the law of conservation of energy when heat is included as a form of energy.

law explains the familiar fact that heat flows from hot to cold (for example, steam to ice) and not the other way around. That being said, heat *can* pass from cold to hot if some energy is expended. Running a heat engine backwards – *spending* energy to pump heat from cold to hot – is the basis of the refrigerator, one of the more lucrative inventions of the Industrial Revolution because it allowed meat to be frozen and transported over thousands of miles.

To understand how Maxwell's demon comes into this, imagine a rigid box containing a gas that is hotter at one end than the other. At the micro-level, heat energy is none other than the energy of motion – the ceaseless agitation of molecules. The hotter the system, the faster the molecules move: at the hot end of the box, the gas molecules move faster, on average, than they do at the cooler end. When the faster-moving molecules collide with the slower-moving ones they will (again, on average) transfer a net amount of this kinetic energy to the cooler gas molecules, raising the temperature of the gas. After a while the system will reach thermal equilibrium, settling down at a uniform temperature partway between the original high and low temperature extremes of the gases. The second law of thermodynamics forbids the reverse process: the gas spontaneously rearranging its molecules so the fast-moving ones congregate at one end of the box and the slow-moving ones at the other. If we saw such a thing, we would think it miraculous.

Although the second law of thermodynamics is easy to understand in the context of boxes of gas, it applies to all physical systems, and indeed to the entire cosmos. It is the second law of thermodynamics that imprints on the universe an arrow of time (see Box 3). In its most general form, the second law is best understood using a quantity called *entropy*. I shall be coming back to entropy in its various conceptions again and again in this story, but for now think of it as a measure of the disorder in a system. Heat, for example, represents entropy because it describes the chaotic agitation of molecules; when heat is generated, entropy rises. If the entropy of a system seems to decrease, just look at the bigger picture and you will find it going up somewhere else. For example, the entropy inside a refrigerator goes down but heat comes out of the back and raises the entropy of the

---

Box 3: Entropy and the arrow of time

Imagine taking a movie of an everyday scene. Now play it backwards; people laugh, because what they see is so preposterous. To describe this pervasive arrow of time, physicists appeal to a concept called *entropy*. The word has many uses and definitions, which can lead to confusion, but the most convenient for our purposes is as a measure of *disorder* in a system with many components. To take an everyday example, imagine opening a new pack of cards, arranged by suit and in numerical order. Now shuffle the cards; they become less ordered. Entropy quantifies that transformation by counting the number of ways systems of many parts can be disordered. There is only *one* way a given suit of cards can be in numerical order (Ace, 2, 3 ... Jack, Queen, King), but *many* different ways it can be disordered. This simple fact implies that randomly shuffling the cards is overwhelmingly likely to increase the disorder – or entropy – because there are so many more ways to be untidy than to be tidy. Note, however, that this is a statistical argument only: there is an exceedingly tiny but non-zero probability that shuffling a suit of jumbled-up cards will accidentally end up placing them in numerical order. Same thing with the box of gas. The molecules are rushing around randomly so there is a finite probability – an exceedingly small probability, to be sure – that the fast molecules will congregate in one end of the box and the slow ones in the other. So the accurate statement is that in a closed system the entropy (or degree of disorder) is *overwhelmingly likely*, but not absolutely certain, to go up, or stay the same. The maximum entropy of a gas – the macroscopic state that can be achieved by the largest number of indistinguishable arrangements – corresponds to thermodynamic equilibrium, with the gas at a uniform temperature and density.

---

kitchen. Added to that, there is a price to be paid in electricity bills. That electricity has to be generated, and the generation process itself makes heat and raises the entropy of the power station. When the books are examined carefully, entropy always wins. On a cosmic scale, the second law implies that *the entropy of the universe never goes down.**

By the middle of the nineteenth century the basic principles of heat, work and entropy and the laws of thermodynamics were well established. There was great confidence that, at last, heat was understood, its properties interfacing comfortably with the rest of physics. But then along came the demon. In a simple conjecture, Maxwell subverted this new-found understanding by striking at the very basis of the second law.

Here is the gist of what was proposed in the letter to Tait. I mentioned that gas molecules rush around, and the hotter the gas, the faster they go. But not all molecules move with the *same* speed. In a gas at a fixed temperature energy is shared out randomly, not uniformly, meaning that some molecules move more quickly than others. Maxwell himself worked out precisely how the energy was distributed among the molecules – what fraction have half the average speed, twice the average, and so on. Realizing that even in thermodynamic equilibrium gas molecules have a variety of speeds (and therefore energies), Maxwell was struck by a curious thought. Suppose it were possible, using some clever device, to separate out the fast molecules from the slow molecules without expending any energy? This sorting procedure would in effect create a temperature difference (fast molecules over here, slow ones over there), and a heat engine could be run off the temperature gradient to perform work. Using this procedure, one would be able to start with a gas at a uniform temperature and convert some of its heat energy into work without any external change, in flagrant violation of the second law. It would in effect reverse the arrow of time and open the way to a type of perpetual motion.

So far, so shocking. But before throwing the rule book of nature into the waste bin, we need to confront the very obvious question of

---

* The great British astronomer Sir Arthur Eddington once wrote: 'The law that entropy always increases, holds, I think, the supreme position among the laws of Nature. If someone points out to you that your pet theory of the universe is in disagreement with Maxwell's [electromagnetic field] equations – then so much the worse for Maxwell's equations. If it is found to be contradicted by observation – well, these experimentalists do bungle things sometimes. But if your theory is found to be against the second law of thermodynamics I can give you no hope; there is nothing for it but to collapse in deepest humiliation.' (Arthur Eddington, *The Nature of the Physical World* (Cambridge University Press, 1928), p. 74) how the separation of fast and slow molecules might actually be attained. Maxwell's letter outlined what he had in mind to accomplish this goal. The basic idea is to divide the box of gas into two halves with a rigid screen in which there is a very small hole (see Fig. 3). Among the teeming hordes of molecules bombarding the screen there will be a handful that arrive just where the hole is located. These molecules will pass through into the other half of the box; if the hole is small enough, only one molecule at a time will traverse it. Left to itself, the traffic in both directions will average out and the temperature will remain stable. But now imagine that the hole could be blocked with a moveable shutter. Furthermore, suppose there were a tiny being – a demon – stationed near the hole and capable of operating the shutter. If it is nimble enough, the demon could allow only slow-moving molecules to pass through the hole in one direction and only fast-moving molecules to go in the other. By continuing this sorting process for a long time, the demon would be able to raise the temperature on one side of the screen and lower it on the other, thus creating a temperature difference without appearing to expend any energy:* order out of molecular chaos, for free.

To Maxwell and his contemporaries the very idea of a manipulative demon violating what was supposed to be a law of nature – that entropy never decreases – seemed preposterous. Clearly, something had been left out of the argument, but what? Well, how about the fact that there are no demons in the real world? That's not a problem. Maxwell's argument falls into the category of what are known as 'thought experiments' – imaginary scenarios that point to some important scientific principles. They don't have to be practical suggestions. There is a long history of such thought experiments in physics and they have frequently led to great advances in understanding, and eventually to practical devices. In any case, Maxwell didn't

---

* Maxwell assumed that the demon and the shutter are perfectly functioning devices with no friction or need for a power source. This is admittedly an idealization, but there is no known principle preventing an arbitrarily close approach to such mechanical perfection. Remember, friction is a macroscopic property where ordered motion, e.g. a ball rolling along the floor, is converted to disordered motion – heat – in which the ball's energy is dissipated among trillions of tiny particles. But on a molecular scale, all is tiny. Friction doesn't exist. Later I will describe some practical demonics.

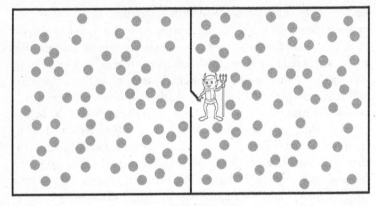

Fig. 3. A box of gas is divided into two chambers by a screen with a small aperture through which molecules may pass one by one. The aperture can be blocked with a shutter. A tiny demon observes the randomly moving molecules and operates the shutter to allow fast molecules to travel from the left-hand chamber to the right-hand one, and slow molecules to go the other way. After a while, the average speed of the molecules on the right will become significantly greater than that on the left, implying that a temperature difference has been established between the two chambers. Useful work may then be performed by running a motor off the heat gradient. The demon thus converts disorganized molecular motion into controlled mechanical motion, creating order out of chaos and opening the way to a type of perpetual motion machine.

need an actual sentient being to operate the shutter, just a molecular-scale device able to perform the sorting task. At the time he wrote to Tait, Maxwell's proposal was a flight of fancy; he can have had no inkling that demonic-type entities really exist. In fact, they existed inside his own body! But the realization of a link between molecular demons and life lay a century in the future.

Meanwhile, apart from the objection 'show me a demon!', nothing else seemed terribly wrong with Maxwell's argument, and for many decades it lay like an inconvenient truth at the core of physics, an ugly paradox that most scientists chose to ignore. With the benefit of hindsight, we can now see that the resolution of the paradox had been lying in plain sight. To operate effectively in sorting the molecules into fast and slow categories, the demon must gather *information* about their speed and direction. And as it turns out, bringing information into physics cracks open a door to a scientific revolution that is only today starting to unfold.

MEASURING INFORMATION

Before I get to the big picture I need to drill down a bit into the concept of information. We use the word a lot in daily life, in all sorts of contexts, ranging from bus timetables to military intelligence. Millions of people work for information technology companies. The growing field of bioinformatics attracts billions of dollars of funding. The economy of the United States is based in large part on information-based industries, and the field is now so much a part of everyday affairs that we usually simply refer to it as 'IT'. But this casual familiarity glosses over some deep conceptual issues. For a start, what exactly *is* information? You can't see it, touch it or smell it, yet it affects everyone: after all, it bankrolls California!

As I remarked, the idea of information derives originally from the realm of human discourse; I may 'inform' students about their exam results, for example, or you may give me the information I need to find the nearest restaurant. Used in that sense, it is a purely abstract concept, like patriotism or political expediency or love. On the other hand, information clearly plays a *physical* role in the world, not least in biology; a change in the information stored in an organism's DNA may produce a mutant offspring and alter the course of evolution. *Information makes a difference in the world.* We might say it has 'causal power'. The challenge to science is to figure out how to couple abstract information to the concrete world of physical objects.

To make progress on these profound issues it is first necessary to come up with a precise definition of information in its raw, unembellished sense. According to the computer on which I am typing this book, the C drive can store 237Gb of information. The machine claims to process information at the rate of 3GHz. If I wanted more storage and faster processing, I would have to pay more. Numbers like this are bandied about all the time. But what are Gb and GHz anyway? (Warning: this section of the book contains some elementary mathematics. It's the only part that does.)

Quantifying information began in earnest with the work of the engineer Claude Shannon in the mid-1940s. An eccentric and somewhat reclusive figure, Shannon worked at Bell Labs in the US, where his primary concern was how to transmit coded messages accurately. The project began as war work: if you are cursed with a hissing radio or a crackly telephone line, what is the best strategy you can adopt to get word through with the least chance of error? Shannon set out to study how information can be encoded so as to minimize the risk of garbling a message. The project culminated in 1949 with the publication of *The Mathematical Theory of Communication.*[2] The book was released without fanfare but history will judge that it represented a pivotal event in science, one that goes right to the heart of Schrödinger's question 'What is life?'

Shannon's starting point was to adopt a mathematically rigorous definition of information. The one he chose turned on the notion of uncertainty. Expressed simply, in acquiring information, you are learning something you didn't know before: ergo, the less uncertain you are about that thing. Think of tossing a fair coin; there is a 50–50 chance as to whether it will land heads or tails. As long as you don't look when it lands, you are completely uncertain of the outcome. When you look, that uncertainty is reduced (to zero, in this example). Binary choices like heads or tails are the simplest to consider and are directly relevant to computing because computer code is formulated in terms of binary arithmetic, consisting only of 1s and 0s. The physical implementation of these symbols requires only a two-state system, such as a switch that may be either on or off. Following Shannon, a 'binary digit', or 'bit', for short, became the standard way of quantifying information. The byte, incidentally, is 8 bits (2^3) and is the b used in Gb (gigabyte, or 1 billion bytes). The speed of information processing is expressed in GHz, standing for 'giga-Hertz', or 1 billion bit flips per second. When you look at the outcome of a fair-coin toss you gain one bit of information by collapsing two equally probable states into one certain state.

What about tossing two coins at once? Inspecting the outcome yields two units of information (bits). Note, however, that when you have two coins there are now *four* possible states: heads-heads, heads-tails, tails-heads and tails-tails. With three coins there are eight possible states and three bits are gained by inspection; four coins have sixteen states and four bits gained; five have thirty-two states . . .; and so on. Notice how this goes: $4 = 2^2$, $8 = 2^3$, $16 = 2^4$, $32 = 2^5$ . . . The number of states is 2 raised to the *power* of the number of coins. Conversely, if you want the number of bits gained by observing the outcome of the coin tosses, this formula must be inverted using *logarithms* to base 2. Thus $2 = \log_2 4$, $3 = \log_2 8$, $4 = \log_2 16$, $5 = \log_2 32$ . . . Those readers familiar with logarithms will notice that this formula makes bits of information additive. For example, 2 bits + 3 bits = 5 bits because $\log_2 4 + \log_2 8 = \log_2 32$, and there are indeed thirty-two equally probable states of five fair coins.

Suppose now that the states are *not* equally probable – for example, if the coin is loaded. In that case, the information gained by inspecting the outcome will be less. If the outcome is completely predictable (probability 1), then no additional information is gained by looking – you get zero bits. In most real-world communications, probabilities are indeed not uniform. For example, in the English language the letter *a* occurs with much higher probability than the letter *x*, which is why the board game Scrabble weights letters differently in the scoring. Another example: in English, the letter *q* is always followed by a *u*, which therefore makes the *u* redundant; you get no more bits of information by receiving a *u* following a *q*, so it wouldn't be worth wasting resources to send it in a coded message.

Shannon worked out how to quantify information in the non-uniform-probability cases by taking a weighted average. To illustrate how, let me give a very simple example. Suppose you flip a loaded coin in which heads occurs on average twice as often as tails, which is to say the probability of heads is ⅔ and the probability of tails is ⅓ (probabilities have to add up to 1). According to Shannon's proposal, the number of bits corresponding to heads or tails is simply weighted by their relative probabilities. Thus, the average number of bits of information obtained from inspecting the outcome of tossing this particular loaded coin is $-\frac{2}{3}\log_2\frac{2}{3} - \frac{1}{3}\log_2\frac{1}{3} = 0.92$ bits, which is somewhat less that the one bit it would have been for equally probable outcomes. This makes sense: if you know heads is twice as likely to come up as tails, there is less uncertainty about the outcome than there would be with a fair coin, and so less reduction in uncertainty by making the observation. To take a more extreme example, suppose heads is seven times as probable as tails. The average number of bits of information per coin toss is now only $-\frac{7}{8}\log_2\frac{7}{8} - \frac{1}{8}\log_2\frac{1}{8} = 0.54$ bits. One way of expressing the information content of an answer to a question is as the average degree of surprise in learning the answer. With a coin so heavily loaded for heads, there generally isn't much surprise.*

A moment's reflection shows that Shannon's analysis has an immediate application to biology. Information is stored in DNA using a universal genetic code. The information content of a gene is transmitted to the ribosome via mRNA, whereupon it is decoded and used to construct proteins from sequences of amino acids. However, the mRNA information channel is intrinsically noisy, that is, error prone (see p. 61). The instruction manual of life is therefore logically equivalent to Shannon's analysis of coded information sent through a noisy communication channel.

What does the surprise factor tell us about how much information an organism contains? Well, life is an exceedingly surprising phenomenon† so we might expect it to possess *lots* of Shannon information. And it does. Every cell in your body contains about a billion DNA bases arranged in a particular sequence of the four-letter biological alphabet. The number of possible combinations is 4 raised to the power of 1 billion, which is one followed by about six hundred million zeros. Compare that to the paltry number of atoms in the universe – one followed by about eighty zeros. Shannon's formula for the information contained in this strand of DNA is to take the

---

* The general formula is $n = -\log_2 p$, where n is the number of bits and p is the probability of each state. That must be a number between 0 and 1, hence the need for a minus sign.

† It certainly is. I wrote this section of the book sitting on a beach near Sydney. When I got to the part about surprise, a stray dog walked unannounced over the keyboard. Here is what it had to add to the discussion: 'V tvtgvtvfaal.' I leave it to the reader to evaluate the information content of this canine interjection.

logarithm, which gives about 2 billion bits – more than the information contained in all the books in the Library of Congress. All this information is packed into a trillionth of the volume of a match head. And the information contained in DNA is only a fraction of the total information in a cell. All of which goes to show just how deeply life is invested in information.*

Shannon spotted that his mathematical formula quantifying information in bits is, bar a minus sign, identical to the physicist's formula for entropy, which suggests that information is, in some sense, the opposite of entropy. That connection is no surprise if you think of entropy as ignorance. Let me explain. I described how entropy is a measure of disorder or randomness (see Box 3). Disorder is a *collective* property of large assemblages; it makes no sense to say a single molecule is disordered or random. Thermodynamic quantities like entropy and heat energy are defined by reference to enormous numbers of particles – for example, molecules of gas careering about – and averaging across them without considering the details of individual particles. (Such averaging is sometimes called a 'coarse-grained view'.) Thus, the temperature of a gas is related to the average energy of motion of the gas molecules. The point is that whenever one takes an average some information is thrown away, that is, we accept some ignorance. The average height of a Londoner tells us nothing about the height of a specific person. Likewise, the temperature of a gas tells us nothing about the speed of a specific molecule. In a nutshell:

* The impressive data-storage properties of DNA have created something of a cottage industry among scientists uploading poetry, books and even films into the DNA of microbes (without killing them). Craig Venter pioneered the field by inserting 'watermarks' in his creation, including a pertinent quotation by the physicist Richard Feynman embedded into the customized genome of a microbe that he re-engineered in his lab. More recently, a group of Harvard biologists encoded a digitized version of the famous galloping-horse movie made by Eadweard Muybridge in 1878 (to demonstrate that all four legs could be off the ground simultaneously) and embedded it into the genomes of a population of living *E. coli* bacteria. (See Seth L. Shipman et al., 'CRISPR–Cas encoding of a digital movie into the genomes of a population of living bacteria', *Nature*, vol. 547, 345–9 (2017).) These feats were more than a bit of recreational tinkering; they provide a graphic demonstration of a technology that could pave the way for inserting 'data recording' devices into cells to keep track of vital processes.

information is about what you know, and entropy is about what you don't know.

As I have explained, if you toss a fair coin and look at the outcome you acquire precisely one bit of information. So does that mean every coin 'contains' precisely one bit of information? Well, yes and no. The answer 'the coin contains one bit' assumes that the number of possible states is two (heads or tails). That's the way we normally think about tossed coins, but this additional criterion isn't absolute; it is relative to the nature of the observation and the measurements you choose to make. For example, there is a lot of information in the figure of the 'head' on the heads side of the coin (same goes for the tails side). If you were an enthusiastic numismatist and had no prior knowledge of the country from which the coin came, or the year, then your quantity of relevant ignorance ('Whose image is on the heads side of the coin?') is much greater than one bit: it is perhaps a thousand bits. In making an observation after tossing heads ('Oh, it's King George V on a 1927 British coin'), you acquire a much greater quantity of information. So the question 'How many bits of information does a coin have?' is clearly undefined as it stands.

The same issue arises with DNA. How much information does a genome store? Earlier, I gave a typical answer (more than the Library of Congress). But implicit in this result is that DNA bases come in a four-letter alphabet – A, T, C, G – implying a one in four chance of guessing which particular base lies at a given location on the DNA molecule if we have no other knowledge of the sequence. So measuring an actual base yields 2 bits of information ($\log_2 4 = 2$). However, buried in this logic is the assumption that all bases are equally probable, which may not be true. For example, some organisms are rich in G and C and poor in A and T. If you know you are dealing with such an organism, you will change the calculation of uncertainty: if you guess G, you are more likely to be right than if you go for A. Conclusion: the information gained by interrogating a DNA sequence depends on what you know or, more accurately, on what you don't know. Entropy, then, is in the eye of the beholder.*

---

* To carry this point to the extreme, let me return to the coin example. Every atom of the coin has a position in space. If you were able to measure where every atom is located, the information acquired would be astronomical. If we ignore quantum

The upshot is that one cannot say in any absolute w
information there is in this or that physical system.[3]
possible to say, however, how much information has
by making a measurement: as stated, information is t
the degree of ignorance or uncertainty about the system being meas
ured. Even if the overall degree of ignorance is ambiguous, the
reduction in uncertainty can still be perfectly well defined.

A LITTLE KNOWLEDGE IS A DANGEROUS THING

If information makes a difference in the world, how should we view
it? Does it obey its own laws, or is it simply a slave to the laws that
govern the physical systems in which it is embedded? In other words,
does information somehow transcend (even if it doesn't actually
bend) the laws of physics, or is it merely, to use the jargon, an epi-
phenomenon, riding on the coat-tails of matter? Does information
per se actually do any work, or is it a mere tracer of the causal activity
of matter? Can information flow ever be decoupled from the flow of
matter or energy?

To address these questions we first have to find a link between
information and physical laws. The first hint of such a link was
already there with Maxwell's demon, but it was left as unfinished
business until the 1920s. At that time Leo Szilárd, a Hungarian Jew
living in Berlin, decided to update Maxwell's thought experiment in
a way that made it easier to analyse.* In a paper entitled 'On the

---

mechanics, a particle such as an electron, which for all we know has no size at all (it
is a point), would represent an *infinite* amount of information because it would take
a set of three infinitely long numbers to specify its exact location in three-dimensional
space. And if just one particle has infinite information, the total information content
of the universe will certainly be infinite.

* Like Schrödinger and many others, Szilárd eventually fled Nazi Europe. He travelled to England and then the United States – which was fortunate for the Allies, as
he was involved in early experiments with nuclear fission. It was Szilárd who, fore-
seeing the possibility of a German atomic bomb, persuaded Einstein to sign a joint
letter to President Roosevelt in 1939, urging the US to develop its own nuclear
weapons.

rease of entropy in a thermodynamic system by the intervention of intelligent beings',[4] Szilárd simplified Maxwell's set-up by considering a box containing only a *single* molecule (see Fig. 4). The end walls of the box are placed in contact with a steady external heat source, which causes them to jitter about. When the trapped molecule strikes a jittering wall, energy is exchanged: if the molecule is moving slowly, it will most likely receive a kick from the wall that speeds it up. If the temperature of the external heat source is raised, the walls will shake harder and the molecule will on average end up going even faster, having bounced off the more vigorously fluctuating walls.* Like Maxwell, Szilárd incorporated a demon and a screen in his (admittedly highly idealized) thought experiment, but he did away with the hole and the shutter mechanism. Instead, Szilárd's demon can effortlessly lift the screen in and out of the box at the mid-point and by the two end walls (there would need to be slots for this). Furthermore, the screen is free to slide back and forth inside the box (without friction). The entire device is known as Szilárd's engine.

Starting with the screen out, the demon is tasked with determining *which side* of the box the molecule is located in. The demon inserts the moveable screen at the mid-point of the box, dividing it in two. Next comes the key step. When the molecule strikes the screen it gives it a little push. Because the screen is free to move it will recoil and so gain energy; conversely, the molecule will lose energy. Though these little molecular knocks will be small by human standards, they can (theoretically) be harnessed to do useful work by raising a weight. To accomplish this, the weight has to be tethered to the screen on the side of the box that contains the molecule; otherwise the weight will fall, not rise (see Fig. 4c). Because the demon knows where the molecule is located, it also knows which side to attach the tether (the attachment can also, in principle, be done with negligible energy expenditure). Thus armed with this modicum of knowledge, i.e.

---

* Because thermal fluctuations are random, the molecule will often move slower or faster than average, but one could imagine a large ensemble of identical single-occupant boxes, and, once equilibrium had been established between the molecular inhabitants of the boxes and the heat source, the distribution of velocities of the ensemble of molecules would precisely mirror those of a gas at the same temperature.

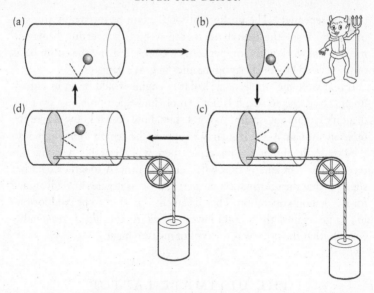

Fig. 4. Szilárd's engine. A box contains a single gas molecule, which can be found in either the right or the left part of the box. (a) Initially, the position of the molecule is unknown. (b) The demon inserts a partition in the centre of the box, and then observes whether the molecule is in the right or the left. (c) Remembering this information, the demon attaches a weight to the appropriate side of the partition. (If the molecule is in the right, as shown, the demon connects the load to the right of the partition.) (d) The molecule, which is moving at high speed due to its thermal energy, collides with the partition, driving it to the left and lifting the weight. In this manner, the demon has converted the random energy of heat into ordered work by using information about the location of the molecule.

positional information, the demon succeeds in converting some of the random thermal energy of the molecule into directed useful work. The demon can wait until the screen has been driven all the way to the end of the box, at which point it can detach the tether, lock the weight in place and slide the screen out of the box at the end slot (all steps that, again in principle, require no energy). The molecule can readily replenish the energy it expended in raising the weight when it collides again with the jittering walls of the box. The entire cycle may then be repeated.* The upshot will once again be the steady transfer of energy from the heat bath to the weight, converting heat into mechanical work with 100 per cent efficiency, placing the entire basis of the second law of thermodynamics in grave danger.

If that were the whole story, Szilárd's engine would be an inventor's dream. Needless to say, it is not. An obvious question hangs over the demon's remarkable faculties. For a start, how does it know where the molecule is? Can it see? If so, how? Suppose the demon shines light into the box to illuminate the molecule; there will inevitably be some unrecoverable light energy that will end up as heat. A rough calculation suggests that the information-garnering process negates any advantage for the demon's operation. There is an entropy price to be paid for trying to go against the second law, and Szilárd concluded, reasonably enough, that the price was the cost of measurement.

THE ULTIMATE LAPTOP

And there the matter might have rested, had it not been for the emergence of a completely different branch of science – the computer industry. While it is true that the demon has to acquire the information about the location of the molecule, that's just the first step. That information has to be *processed* in the demon's diminutive brain, enabling it to make decisions about how to operate the shutter in the appropriate manner.

When Szilárd invented his engine, information technology and computers lay more than two decades in the future. But by the 1950s

---

* Using Shannon's formula, I can be more precise about the demon. In Szilárd's engine, the molecule is equally likely to be on the left of the box or on the right. By observing which, the demon reduces the uncertainty from fifty:fifty to zero, thus (according to Shannon's formula) acquiring precisely one bit of information. When this solitary bit is used to lift a weight, the amount of energy extracted from the heat reservoir depends on the temperature, T. It is a simple calculation to work out the force on the screen due to its bombardment by the confined molecule; the higher the temperature, the greater the force. From that calculation one soon finds that the theoretical maximum amount of work extracted by Szilárd's engine is $kT \ln 2$. Here the quantity k (known as Boltzmann's constant) is needed to express the answer in units of energy, such as joules. Putting in the numbers, at room temperature, one bit of information yields 3×10^{-21} joules.

general purpose digital computers of the sort we are familiar with today (such as the one on which I am typing this book) were advancing in leaps and bounds. A leading business propelling this effort was IBM. The company set up a research facility in upstate New York, recruited some of the brightest minds in mathematics and computing and charged them with the task of discovering 'the laws of computing'. Computer scientists and engineers were eager to uncover the fundamental principles that constrain exactly what can be computed and how efficiently computing can be done. In this endeavour, the computer scientists were retracing similar steps to nineteenth-century physicists who wanted to work out the fundamental laws of heat engines. But this time there was a fascinating refinement. Because computers are themselves physical devices, the question arises of how the laws of computing mesh with the laws of physics – especially the laws of thermodynamics – that govern computer hardware. The field was ripe for the resurrection of Maxwell's demon.

A pioneer in this formidable challenge was Rolf Landauer, a German-born physicist who also fled the Nazis to settle in the United States. Landauer was interested in the fundamental *physical* limits of computing. It is a familiar experience when using a laptop computer on one's lap that it gets hot. A major financial burden of computing has to do with dissipating this waste heat, for example with fans and cooling systems, not to mention the cost of the electricity to pay for it all. In the US alone, waste heat from computers drains the GDP by $30 billion a year, and rising.*

Why do computers generate heat? There are many reasons, but one of them goes to the very heart of what is meant by the term 'computation'. Take a problem in simple arithmetic, like long division, that can also be carried out with a pencil and paper. You start with two numbers (the numerator and the denominator) and end up with one number (the answer) plus some procedural scribbles needed to get there. If all you are interested in is the answer – the 'output', in computer jargon – then the input numbers and all the intermediate steps can be thrown away. Erasing the preceding steps makes the computation logically

---

* A much-discussed example is bitcoin mining, which is estimated to consume more power than Denmark.

irreversible: you can't tell by looking at the answer what the question was. (Example: 12 might be 6 x 2 or 4 x 3 or 7 + 5.) Electronic computers do the same thing. They take input data, process it, output the answer, and (usually only when memory needs freeing up) irreversibly erase the stored information.

Acts of erasure generate heat. This is familiar enough from the long-division example: removing the pencil workings with a rubber eraser involves a lot of friction, which means heat, which means entropy. Even sophisticated microchips generate heat when they get rid of 1s and 0s.* What if one could design a computer that processed information without producing *any heat at all*? It could be run at no cost: the ultimate laptop![5] Any company that achieved such a feat would immediately reign supreme in the computing business. No wonder IBM was interested. Sadly, Landauer poured cold water on this dream by arguing that when the information processed in a computer involves logically irreversible operations (as in the arithmetic example above), there will inevitably be heat dissipated when the system is reset for the next computation. He calculated the minimum amount of entropy needed to erase one bit of information, a result now known as the Landauer limit. For the curious, erasing one bit of information at room temperature generates 3×10^{-21} joules, about a hundred trillion trillionth of the heat energy needed to boil a kettle. That's not a lot of heat, but it establishes an important principle. By demonstrating a link between logical operations and heat generation, Landauer found a deep connection between physics and information, not in the rather abstract demonic sense of Szilárd, but in the very specific (that is, dollar-related) sense in which it is understood in today's computing industry.[6]

From Landauer on, information ceased to be a vaguely mystical quantity and became firmly anchored in matter. To summarize this transformation in thinking, Landauer coined a now-famous dictum: 'Information is physical!'[7] What he meant by this is that all information must be tied to physical objects: it doesn't float free in the ether. The information in your computer is stored as patterns on the hard drive, for example. What makes information a slippery concept is

---

* There is an important difference between physically eliminating the 1s and 0s and resetting the state of the computer's memory to some reference state, such as all 0s, creating a 'tabula rasa'. It was the latter that Landauer studied.

that the particular physical instantiation (the actual substrate) often doesn't seem important. You can copy the contents of your hard drive onto a flash drive, or relay it via Bluetooth, or send it in laser pulses down a fibre or even into space. So long as it is done properly, the information stays the same when it is transferred from one variety of physical system to another. This independence of substrate seems to give information 'a life of its own' – an autonomous existence.

In this respect, information shares some of the properties of energy. Like information, energy can be passed from one physical system to another and, under the right conditions, it is conserved. So would one say that energy has an autonomous existence? Think of a simple problem in Newtonian mechanics: the collision of two billiard balls. Suppose a white ball is skilfully propelled towards a stationary red ball. There is a collision and the red ball flies off towards a pocket. Would it be accurate to say that 'energy' caused the red ball to move? It is true that the kinetic energy of the white ball was needed to propel the red ball, and some of this energy was passed on in the collision. So, in that sense, yes, energy (strictly, energy transfer) was a causative factor. However, physicists would not normally discuss the problem in these terms. They would simply say that the white ball hit the red ball, causing it to move. But because kinetic energy is instantiated in the balls, where the balls go, the energy goes. So to attribute causal power to energy isn't wrong, but it is somewhat quixotic. One could give a completely detailed and accurate account of the collision without any reference to energy whatsoever.

When it comes to information, are we in the same boat? If all causal power is vested in the underlying matter in which information is instantiated, it might be regarded as equally quixotic, albeit convenient, to discuss information as a cause. So is information real, or just a convenient way to think about complex processes? There is no consensus on this matter, though I am going to stick my neck out and answer yes, information does have a type of independent existence and it does have causal power. I am led to this point of view in part by the research I shall describe in the next chapter involving tracking shifting patterns of information in networks that do indeed seem to obey certain universal rules transcending the actual physical hardware in which the bits are instantiated.

READING THE MIND OF THE DEMON

If Landauer's limit is truly fundamental, then it must also apply to the information processed in the demon's brain. Landauer didn't pursue that line of inquiry himself, however. It took another IBM scientist, Charles Bennett, to investigate the matter, twenty years later. The prevailing view was still that the demon can't violate the second law of thermodynamics because any entropy-reducing advantage gained by its antics is negated by the entropy-generating cost of sensing the molecules in the first place. But reflecting deeply on this matter led Bennett to suspect there was a flaw in the received wisdom. He worked out a way to detect the state of a molecule without generating any entropy at all.* If the second law is to be saved, Bennett argued, then the compensating entropy cost must come from somewhere else. At first sight there was a ready answer: the irreversible nature of computation – the merging of numbers needed to output an answer. That would certainly produce heat if carried out directly. But even here Bennett found a loophole. He pointed out that *all* computations can in fact be made reversible. The idea is simple. In the pencil-and-paper example I gave it would be enough to merely keep a record of the input and all the intermediate steps in order to run the long division backwards. You could easily begin with the answer and finish by outputting the question, because everything you need is there on the paper. In an electronic computer the same thing can be achieved with specially designed logic gates, wired together to form circuits in such a way that all the information is retained somewhere in the system. With this set-up, no bits are erased and no heat is produced; there is no rise in entropy. I should stress that today's computers are very far indeed from the theoretical possibility of reversible computation. But we are dealing here with deep issues of principle, and there is no known reason why the theoretical limit may not one day be approached.

Now we are back to square one as far as the demon is concerned. If it can acquire information about molecules at negligible entropic cost, process the information reversibly in its tiny brain and

---

* His examples of how to do it were highly idealized and need not concern us here.

effortlessly operate a shutter, then by repeating the process again and again the demon would be able to generate perpetual motion.

What's the catch? There is one, and according to Bennett it is buried in the 'again and again' qualification.[8] Let's take stock: the demon has to process the acquired information to operate the mechanism correctly. The processing could, in principle, be carried out reversibly, producing no heat, but only if the demon retains all the intermediate steps of the computation in its memory. Fine. But if the demon repeats the trick, more information will be added, and yet more on the next cycle, and so on. Over time, the demon's internal memory will inexorably clog up with bits of information. So the sequence of computations can all be reversed so long as there is enough memory space. To operate in a truly open-ended manner a finite demon needs to be brainwashed at the end of each cycle; that is, the demon's memory must be cleared and its state reset to the one it had at the outset before it embarks on the next cycle. And this step proved to be the demon's Achilles heel. Bennett proved that the *act of information erasure* generates just the right amount of entropy to pay for the apparent violation of the second law attained by the demon's activities.

Nevertheless, the subject of the demon continues to attract dissent and controversy. For example, what happens if one has an endless supply of demons, so when one gets its brain clogged, another is substituted? Also, a more general analysis suggests that a demon could be made in which the sum of the entropy of observation and erasure entropy can never be less than the Landauer bound. In this system, the entropic burden can be distributed between observation and erasure in any mix.[9] Many open questions remain.

INFORMATION ENGINES

The way I've described it, there's still something a bit magical about how the demon operates as an intelligent agent. Surely it doesn't have to be sentient, or even intelligent in the everyday IQ sense? It must be possible to substitute a mindless gadget – a demonic automaton – that would serve the same function. Recently, Christopher Jarzynski at the University of Maryland and two colleagues dreamed up such a

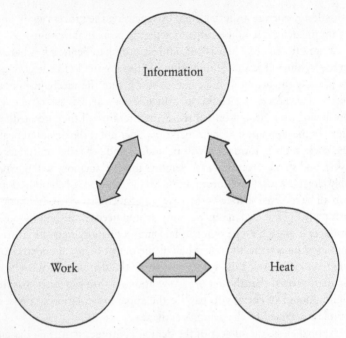

Fig. 5. The three-way trade-off of information, heat energy and work. Maxwell's and Szilárd's demons process information to convert heat into work. Information engines do work by turning information into heat or by dumping entropy into an empty information register. Conventional engines use heat to do work and thereby destroy information (i.e. create entropy).

gadget, which they call an information engine. Here is its job description: 'it systematically withdraws energy from a single thermal reservoir and delivers that energy to lift a mass against gravity while writing information to a memory register'.[10] Though hardly a practical device, their imaginary engine provides a neat thought experiment to assess the three-way mix of heat, information and work, and to help us discover their relative trade-offs.

The Jarzynski contraption resembles a child's plaything (see Fig. 6). The demon itself is simply a ring that can rotate in the horizontal plane. A vertical rod is aligned with the axis of the ring, and attached to the rod are paddles perpendicular to the rod which stick out at

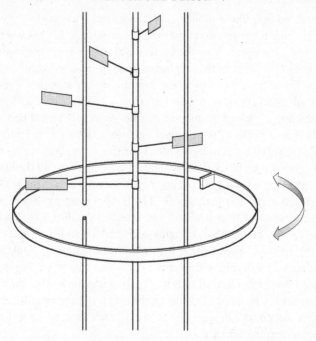

Fig. 6. Design for an information engine. In this variant of Maxwell's demon experiment, the central rod descends continuously through the ring. Two fixed vertical rods lie on either side of, and are co-planar with, the central rod. One of these has a gap at the level of the ring. Identical paddles attached to the central rod are free to swivel horizontally; their positions encode 0 or 1 respectively, depending on whether they lie on the far side or near side of the three rods as viewed in the figure. In the configuration shown, the initial state consists of all 0s. The horizontal ring serves as a simple demon. It can rotate freely in the horizontal plane. It has a single projecting blade designed so that it can be struck by the swivelling paddle blades sending it either clockwise or anticlockwise. The entire gadget is immersed in a bath of heat and so all components will experience random thermal fluctuations. Because of the gap in the left-hand rod there are more anticlockwise blows to the ring than clockwise (as viewed from above). The device therefore converts random thermal motion into directional rotation that could be used to raise a weight, but in so doing the output state of the blades (their configuration below the ring; none shown here) is now a random mixture of 1s and 0s. The machine has thus converted heat into work and written information into a register.

different angles, like a mobile, and can swivel frictionlessly on the rod. The precise angles don't matter; the important thing is whether they are on the near side or the far side of the three co-planar rods as shown. On the far side, they represent 0; on the near side, they represent 1. These paddles serve as the demon's memory, which is just a string of digits such as 0100101011010 ... The entire gadget is immersed in a bath of heat so the paddles randomly swivel this way and that as a result of the thermal agitation. However, the paddles cannot swivel so far as to flip 0s into 1s or vice versa, because the two outer vertical rods block the way. The show begins with all the blades above the ring set to 0, that is, positioned somewhere on the far side as depicted in the figure; this is the 'blank input memory' (the demon is brainwashed). The central rod and attached paddles now descend vertically at a steady pace, bringing each paddle blade one at a time into the ring and then exiting below it. So far it doesn't look as if anything very exciting will happen. But – and this is an important feature – one of the vertical rods has a gap in it at the level of the ring, so now as each blade passes through the ring it is momentarily free to swivel through 360 degrees. As a result, each descending 0 has a chance of turning into a 1.

Now for the crucial part. For the memory to be of any use to the demon, the descending blades need to somehow interact with it (remember that, in this case, the demon is the ring) or the demon cannot access its memory. The interaction proposed by the designers is very simple. The demonic ring comes with a blade of its own which projects inwards and is fixed to the ring; if one of the slowly descending paddles swivels around in the right direction its blade will clonk the projecting ring blade, causing the ring to rotate in the same direction. The ring can be propelled either way but, due to the asymmetric configuration of the gap, there are more blows sending the ring anticlockwise than clockwise (as viewed from above). As a result, the random thermal motions are converted into a cumulative rotation in one direction only. Such progressive rotation could be used in the now familiar manner to perform useful work. For example, the ring could be mechanically coupled to a pulley in such a way that anticlockwise movement of the ring would raise a weight and clockwise movement lower it. On average, the weight will rise. (If all this sounds too complicated to grasp, there is a helpful video animation available.)[11]

So what happened to the second law of thermodynamics? We seem once more to be getting order out of chaos, directed motion from randomness, heat turning into work. To comply with the second law, entropy has to be generated somewhere, and it is: in the memory. Translated into descending blade configurations, some os become 1s, and some os stay os. The record of this action is preserved below the ring, where the two blocking rods lock in the descending state of the paddles by preventing any further swivelling between o and 1. The upshot is that Jarzynski's device converts a simple ordered input state oooooooooooooooo . . . into a complex, disordered (indeed random) output state, such as 1ooot011101001o . . . Because a string of straight os contains no information, whereas a sequence of 1s and os is information rich,* the demon has succeeded in turning heat into work (by raising the weight) and accumulating information in its memory. The greater the storage capacity of the incoming information stream, the larger the mass the demon can hoist against gravity. The authors remark, 'One litre of ordinary air weighs less than half a US penny, but it contains enough thermal energy to toss a 7kg bowling ball more than 3m off the ground. A gadget able to harvest that abundant energy by converting the erratic movement of colliding molecules into directed motion could be very useful indeed.'[12] And so it would. But just like Maxwell's and Szilárd's demons, Jarzynski's demon can't work repetitively without clearing the memory and erasing information, a step that inevitably raises the entropy.

In fact, Jarzynski's engine can be run in reverse to erase information. If instead of straight os, the input state is a mixture of 1s and os (representing information), then the weight descends, and in so doing it pays for the erasure with its own gravitational potential energy. In this case, the output has more os than the input. The designers explain: 'When presented with a blank slate the demon can lift any mass; but when the incoming bit stream is saturated [with random numbers] the demon is incapable of delivering work . . . Thus a blank or partially blank memory register acts as a thermodynamic resource

---

* For example, o could stand for dot and 1 for dash in Morse code.

that gets consumed when the demon acts as an engine.'[13] This is start-ling. If erasing information increases entropy, then acquiring an empty memory amounts to an injection of fuel. In principle, this tab-ula rasa could be anything at all – a magnetic computer memory chip or a row of os on a paper tape. According to Jarzynski, 300 billion billion zeros could lift an apple, demonically, by one metre!

The notion that the *absence* of something (a blank memory) can be a physical resource is reminiscent of the Improbability Drive in *The Hitchhiker's Guide to the Galaxy*.[14] But weird though it seems, it is the inevitable flip side to Charles Bennett's analysis. No doubt the reader is nonplussed at this stage. Can a string of zeros really run an engine? Can information itself serve as a fuel, like petrol? And is this just a collection of mind games, or does it connect to the real world?

DEMONS FOR DOLLARS: INVEST NOW IN APPLIED DEMONOLOGY

One hundred and forty years after Maxwell first floated the idea, a real Maxwell demon was built in the city of his birth. David Leigh and his collaborators at Edinburgh University published the details in a paper in *Nature* in 2007.[15] For over a century the idea of anyone actually building a demon seemed incredible, but such have been the advances in technology – most significantly in nanotechnology – that the field of applied demonology has at last arrived.*

The Leigh group built a little information engine consisting of a molecular ring that can slide back and forth on a rod with stoppers at the end (like a dumbbell). In the middle of the rod is another mol-ecule that can exist in two conformations, one that blocks the ring and one that allows it to pass over the blockage. It thus serves as a gate, similar to Maxwell's original conception of a moveable shutter.

* When the physicist Richard Feynman died, he left a famous aphorism on his blackboard: 'What I cannot build, I cannot understand.' (It was these words that Craig Venter inscribed into his artificial organism – see p. 39.) Today, scientists are building Maxwell demons and information engines, and elucidating their operating principles. The place of information in the eternal tussle between order and chaos is finally being revealed in a practical way.

The gate can be controlled with a laser. The system is in contact with surroundings that are maintained at a finite temperature, so the ring will jiggle randomly back and forth along the rod as a result of normal thermal agitation. At the start of the experiment it is confined to one half of the rod with its movement blocked by the 'gate' molecule set to the 'closed' position. The researchers were able to follow the antics of the ring and gate in detail and test that the system really is driven away, demon-like, from thermodynamic equilibrium. They confirmed that 'information known to a gate-operating demon' serves as a fuel, while its erasure raises entropy 'in agreement with Bennett's resolution of the Maxwell demon paradox'.[16]

The Edinburgh experiment was quickly followed by others. In 2010 a group of Japanese scientists manipulated the thermal agitation of a tiny polystyrene bead and announced, 'We have verified that information can indeed be converted to potential energy and that the fundamental principle of the demon holds true.'[17] The experimenters reported that they were able to turn information into energy with 28 per cent efficiency. They envisaged a future nano-engine that runs solely on 'information fuel'.

A third experiment, performed by a group at Aalto University in Finland, got right down to the nano-scale by trapping a single electron in a tiny box just a few millionths of a metre across held at a low but finite temperature. The electron started out free to visit one of two locations – just as with the box in Szilárd's engine. A sensitive electrometer determined where the electron resided. This positional information was then fed into a device that ramped up the voltage (a reversible operation with no net energy demand) so as to trap the electron in situ – analogous to the demon inserting the screen. Next, energy was slowly extracted from the electron's thermal movement and used to perform work. Finally, the voltage was returned to its starting value, completing the cycle. The Finnish team carried out this experiment 2,944 times, attaining an average of 75 per cent of the thermodynamic limit of a perfect Szilárd engine. Importantly, the experiment is an *autonomous* Maxwell demon, 'where only information, not heat, is directly exchanged between the system and the demon'.[18] The experimenters themselves didn't meddle in the process and, indeed, they didn't even know where the electron was each time – the measurement and feedback control activity were entirely automatic and self-contained: no external agent was involved.

In a further refinement, the Finnish team coupled two such devices together, treating one as the system and the other as the demon. Then they measured the demonically extracted heat energy by monitoring the cooling of the system and the corresponding heating up of the demon. They touted this nanotechnological feat as the creation of the world's first 'information-powered refrigerator'. Given the pace of technological advancement, demonic devices of this sort will likely become available by the mid-2020s.* Expect a big impact on the commercialization of nanotechnology, but probably a smaller impact on kitchen appliances.

ENGINES OF LIFE: THE DEMONS IN YOUR CELLS

'Information is the currency of life.'

Christoph Adami[19]

Though Maxwell would doubtless have been delighted to see the advent of practical demonology, he could scarcely have guessed that the interplay of information and energy involved has been exploited by living organisms for billions of years. Living cells, it turns out, contain a host of exquisitely efficient and well-honed nano-machines, made mostly from proteins. The list includes motors, pumps, tubes, shears, rotors, levers and ropes – the sort of paraphernalia familiar to engineers.

Here is one amazing example: a type of turbine consisting of two aligned rotors coupled by a shaft. (Its function in living cells is to play a role in energy transport and storage.) The rotor turns when protons

---

* In a related experiment on the trade-off between information and heat flow, performed as part of the Brazilian National Institute of Science and Technology for Quantum Information programme, it was reported that heat was induced to flow from a colder to a hotter system (i.e. refrigeration) by using entangled quantum particles, a subject I shall explain in Chapter 5.

(there are always plenty of them roaming around inside cells) traverse the shaft in one direction. If the rotor is run backwards, it pumps out protons in the reverse direction. A Japanese group conducted an ingenious experiment in which one of the rotors was extracted and anchored to a glass surface for study. They attached a molecular filament to the end of the shaft and tagged it with a fluorescent dye so it could be seen under a light microscope when a laser was shone on it. They were able to watch the rotor turn in discrete jumps of 120 degrees each time a proton transited.[20]

Another tiny biomachine that has attracted a lot of attention is a sort of freight-delivery molecule called kinesin. It carries vital cargoes by walking along the tiny fibres which criss-cross cells. It does this gingerly – one careful step at a time – to avoid being swept away by the incessant bombardment from the thermally agitated water molecules that saturate all living cells and move twice as fast as a jetliner. One foot stays anchored to the fibre and the other comes from behind and sets down ahead; then the process is repeated with the other foot. The anchor points are where the binding forces between the foot and the fibre are especially propitious: those sites are 8 nanometres apart, so each step is 16 nanometres in length. It's unnerving to think that billions of these little kinesin wonders are creeping around inside you all the time. Readers should check out an entertaining YouTube cartoon showing kinesin strutting its stuff.[21] (Those who want more technical details should read Box 4.)

An obvious question is what makes this mindless molecular machine exhibit patently purposeful progress? If it simply lifts a foot, then thermal agitation will propel it forwards and backwards at random. How does it plough doggedly ahead in the teeth of the relentless molecular barrage? The answer lies in the way that kinesin acts as a form of ratchet (one foot always anchored, remember). Molecular ratchets are a good example of demons, which are basically in the business of using information to convert random thermal energy into directional motion.* But, to avoid falling foul of the second law, kinesin must tap into a power source.

---

* A process known as rectification.

Box 4: How kinesin can walk the walk

ATP – life's miracle fuel – is converted into a related molecule called ADP (adenosine diphosphate) when it gives up its energy. ADP can be 'recharged' to ATP, so ATP is recycled rather than discarded when it has delivered its energy. ATP and ADP are critical to the operation of the kinesin walker. The kinesin has a little socket in the 'heel' of each foot shaped exactly so that an ADP molecule can fit into it snugly and bind to it. When the slot is thus occupied, the shape of the leg changes a little, causing the foot to detach from the fibre, when it is then free to move around. When the loose foot locates the next anchor point it releases the ADP from its slot, causing the foot to bind once more to the fibre. While this foot-loose activity is going on, the other foot (the one initially in front) had better hang onto the fibre: if both feet came free together, the kinesin molecule would drift away and its cargo would be lost. The other foot – now the back foot – will stay anchored so long as its own ADP slot remains empty. But will it? Well, the very same heel slot that binds ADP can also accommodate ATP. If a randomly passing ATP encounters the empty slot of the anchored back foot, it will snap into it. Then three things happen. First, the kinesin molecule deforms and reorients in such a way as to frustrate any attempt by passing ATPs to fill the now-empty slot of the front foot. Second, ATP contains stored chemical energy. In the slot it undergoes a chemical transformation ATP → ADP and thereby releases its energy into the little kinesin machine. The resulting kick contributes to driving the machine, but also – thirdly – the conversion to ADP means that the back foot now contains an ADP molecule in its slot, as a result of which it detaches from the fibre and begins the process of walking forward so the cycle can be repeated.[22]

Let me digress a moment to explain the energetics here, as it is important more generally. Biology's fuel of choice is a molecule called ATP (adenosine triphosphate); it's like a mini-power-pack with a lot of punch and it has the useful feature that it can keep its energy stored until needed, then – *kerpow!* Life is so fond of ATP fuel for its myriad nano-machines (like the abovementioned rotor, for example) it's been estimated that some organisms burn through their entire body weight of the stuff in just one day.

Life uses many ratchets. The kinesin walker is one example designed to go forwards, not forwards and backwards equally. Looking at the physics of ratchets subject to thermal fluctuations leads to a clear conclusion. They work only if there is either a source of energy to drive them in one direction or active intervention by an information-processing system (demon). No fuel, or no demon, means no progress. Entropy is always generated: in the former case from the conversion of the driving energy into heat; in the latter from the entropy of information processing and memory erasure. There is no free lunch. But by ratcheting ahead instead of simply 'jet-packing' its cargo through the molecular barrage, the lunch bill for kinesin is greatly reduced.

Box 5: Feynman's ratchet

An attempt to replace Maxwell's demon by a purely passive device was made by Richard Feynman. He considered a ratchet of the sort employed by mechanical clockwork (needed so the hands don't turn anticlockwise; see Fig. 7). It involves a toothed wheel with a spring-loaded pawl to stop the wheel slipping backwards. Critical to the operation of the ratchet is the asymmetry of the teeth: they have a steep side and a shallow side. This asymmetry defines the direction of rotation; it is easier for the pawl to slide up the shallow edge of a tooth than the steep edge. Feynman then wondered whether, if the ratchet were immersed in a heat bath maintained at uniform temperature, random thermal fluctuations might occasionally cause the wheel to advance in the forward direction (clockwise in the diagram) but not in the reverse direction. If the ratchet were attached to a rope, it would be able to lift a weight, thus doing useful work powered only by heat. Not so. The flaw in the argument rests with the spring-loaded pawl. In thermodynamic equilibrium it too will jiggle about due to thermal fluctuations, sometimes causing the wheel to slip back the wrong way. Feynman calculated the relative probabilities of forward and reverse motion of the ratchet wheel and argued that, on average, they balanced out.[26]

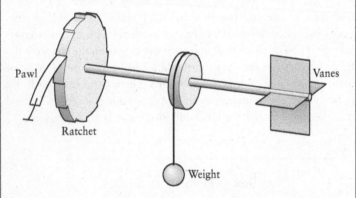

Fig. 7. Feynman's ratchet. In this thought experiment gas molecules bombard the vanes, causing the shaft to rotate randomly clockwise or anticlockwise. If it jerks clockwise, the ratchet permits the shaft to turn, thus raising the weight. But if the shaft tries to rotate anticlockwise, the pawl blocks it. Thus the device seems to convert the heat energy of the gas into work, in violation of the second law of thermodynamics.

Now for an arresting postscript. There's another make of walker called dynein that, in what seems like a fit of designer madness, walks *the other way* along the same fibres as kinesin. The inevitable encounters between kinesin and dynein lead to occasional episodes of road rage and call for some deft manoeuvring. There are even road blocks stationed along the fibres requiring side-stepping and other molecular dances. Yet biology has solved all these problems with remarkable ingenuity: by exploiting demonic ratcheting, kinesin operates at an impressive 60 per cent efficiency in regard to its ATP fuel consumption. (Compare a typical car engine, which is about 20 per cent efficient.)

The core of life's machinery revolves round DNA and RNA, so it's no surprise that nature has also honed the minuscule machines that attend them to also operate at high thermodynamic efficiency. One example is an enzyme known as RNA polymerase, a tiny motor whose job is to crawl along DNA and copy (transcribe) the digital information into RNA, letter by letter. The RNA strand grows as it goes, adding matching letters at each step. It turns out that this mechanism comes very close to the theoretical limit of being a Maxwell demon, consuming almost no energy. We know it isn't quite 100 per cent accurate because there are occasional errors in the transcript (which is good: remember, errors are the drivers of Darwinian evolution). Errors can be corrected, however, and they mostly are. Life has devised some amazingly clever and efficient ways to read the RNA output and fix up goofs.* But in spite of all this striving to get it right, there is a very basic reason why RNA error correction can't be perfect: there are many ways for the transcription to be wrong but only one way for it to be right. As a result, error-correction is irreversible; you can't infer the erroneous sequence from the corrected sequence. (This is another example of not being able to deduce the question from the answer.) Logically, then, the error-correcting process merges many distinct input states into a single output state, which, as we know from Landauer's work, always carries an entropy cost (see p. 46).

A different demonic motor springs into action when the cell divides and DNA is duplicated. Called DNA polymerase, its job is to copy from one DNA strand into another, daughter molecule, which again is built one letter at a time as the motor crawls along. It typically moves at about 100 base pairs per second and, like RNA polymerase, it also operates at close to thermodynamic perfection. In fact, it is possible to run this mechanism in reverse by the simple expedient of

---

* Like all physical processes, even this fix occasionally goes wrong. In humans, for example, the proofread and edited RNA transcript still contains about one mistake in every 100 million letters.

tensioning the DNA. In the lab this can be done with devices called optical tweezers. As the tension increases, so the enzyme crawls more slowly until, at a tension of about 40pN it stops completely. (pN here stands for 'pico-newton', or a trillionth of a newton, a standard unit of force named after the great man himself.) At higher tensions, the tiny motor goes backwards and undoes its handiwork letter by letter.[23]

Copying DNA is of course only a small part of the process of reproduction whereby a cell divides into two. An interesting engineering question is how much the whole cell-reproduction process costs, energy/entropy-wise. Jeremy England of MIT analysed this topic with bacteria,[24] which hold the world record for rapid reproduction (twenty minutes going flat out). Given what I have explained about heat and entropy, the question arises of whether or not the bacteria grow hot as a result. Well, they do, but not as hot as you might imagine from all that pushing, pulling and molecular rearrangement. According to England, *E. coli* generate only about six times the heat of the theoretical minimum limit set by thermodynamics, so they are almost as efficient on the cellular level as they are on the nano-machine level.*

How can we explain the astonishing thermodynamic efficiency of life? Organisms are awash with information, from DNA up to social organization, and it all comes with an entropy cost. No surprise, then, that evolution has refined life's information-management machinery to operate in a super-efficient manner. Organisms need to have perfected the art of storing and processing information or they would quite simply cook themselves to death with waste heat.

Though life's nano-machines, on the face of it, obey the same laws of physics as familiar macroscopic machines, the environment in

---

* You might be wondering how on Earth England could calculate the reproduction entropy, given the complexity of a bacterial cell. He is able to do it because of a fundamental connection between the entropy generated by a physical process and the rate at which the process proceeds compared to the reverse process. For example, the reverse process to synthesizing a strand of RNA from its components is destroying the said strand of RNA in similar circumstances. How might that happen? In water, RNA disintegrates spontaneously in about four days. Compare that to the roughly one hour it takes for it to be synthesized. From this ratio, the theoretical minimum entropy production can be calculated for this piece of the action. Taking all the relevant factors into consideration leads to England's quoted estimate.

which they operate is starkly different. A typical mammalian cell may contain up to 10 billion proteins, which places them on average only a few nanometres apart. Every nano-machine is continually buffeted by the impact of high-speed water molecules, which make up much of the cell's mass. Conditions resemble those in a rowdy nightclub: crowded and noisy. The little machines, unless anchored, will be knocked all over the place. Such mayhem might seem like a problem for the smooth running of the cellular machinery, but it can also be a positive advantage. After all, if the interior of the cell were frozen into immobility, nothing at all would happen. But there is a downside to the incessant thermal disruption: life must expend a lot of effort repairing the damage and rebuilding disintegrating structures.

One way to think about thermal noise is in terms of average energy of molecular motion. Big molecules such as proteins move more slowly than water molecules, but as they are much more massive (a typical protein weighs as much as 10,000 water molecules) they carry about the same amount of energy. Thus, there is a natural unit of energy at any given temperature: at room temperature it is about 3×10^{-21} joules. This will be the energy of a typical molecule. It also happens to be about the same as the energy needed to deform the shapes of important molecular structures like kinesin, and furthermore, that needed to unravel or fracture molecules. Much of life's machinery thus teeters on the edge of heat destruction. Again, this seems like a problem, but it is in fact vital. Life is a process, and the disruption wrought by the unrelenting molecular clamour provides an opportunity for rearrangement and novelty. It also makes the conversion between one form of energy and another easy. For example, some biological nano-machines turn electrical energy into motion; others turn mechanical energy into chemical energy.

The reader might be wondering why so many vital processes take place invisibly, on a nanoscale, under such trying and extreme conditions. The aforementioned coincidence of energy scales provides a ready answer. For life as we know it, liquid water plays a critical role, and that brackets the temperature range at which biology can operate. It turns out to be only at the nano-scale that the thermal energy in this temperature range is comparable to the chemical and mechanical energy of the biological machinery, and thus able to drive a wide range of transformations.[25]

BEYOND THE BIT

Living organisms, we now know, are replete with minuscule machines chuntering away like restless Maxwell demons, keeping life ticking over. They manipulate information in clever, super-efficient ways, conjuring order from chaos, deftly dodging the strictures of thermodynamics' kill-joy second law. The biological information engines I have described, and their technological counterparts, involve simple feedback-and-control loops. Although the actual molecules are complex, the logic of their function is simple: just think of kinesin, tirelessly working away at the 'molecular coalface'.

The cell as a whole is a vast web of information management. Consider, for example, the coded information on DNA. Making proteins is a complicated affair, over and above the mRNA transcription step. Other proteins have to attach the right amino acids to strands of transfer RNA, which then bring them to the ribosome for their cargoes to be hooked together on cue. Once the chain of amino acids is completed, it may be modified by yet other proteins in many different ways, which we'll explore in Chapter 4. It must also fold into the appropriate three-dimensional structure, assisted by yet more proteins that chaperone the flexible molecule during the folding process. All this exquisite choreography has to work amid the thermal pandemonium of the cell.

On its own, the information in the gene is static, but once it is read out – when the gene is *expressed* as the production of a protein – all manner of activity ensues. DNA output is combined with other streams of information, following various complex pathways within the cell and cooperating with a legion of additional information flows to produce a coherent collective order. The cell integrates all this information and progresses as a single unit through a cycle with various identifiable stages, culminating in cell division. And if we extend the analysis to multicelled organisms, involving the astounding organization of embryo development, then we are struck even more forcibly that simply invoking 'information' as a bland, catch-all quantity, like energy, falls far short of an explanation for what is going on.

This is where Shannon's definition of information, important though it is, fails to give a complete account of biological information. It is deficient in two important respects:

1. Genetic information is contextual. Shannon himself was at pains to point out that his work dealt purely with transmitting bits of information defined in the most economical manner and had nothing to say about the *meaning* of the message encoded. The quantity of Shannon information is the same whether a DNA sequence encodes instructions to build a protein or is just arbitrary 'junk' DNA. But the consequences for biological functionality are profound: a protein will fulfil a vital task; junk will do nothing of significance. The difference is analogous to Shakespeare versus a random jumble of letters. For genetic information to attain functionality, there has to be a molecular milieu – a global context – that recognizes the instructions and responds appropriately.

2. Organisms are prediction machines. At the level of the organism as a whole, information is gathered from an unpredictable and fluctuating environment, manipulated internally and an optimal response initiated. Examples include a bacterium swimming towards a source of food and ants exploring their surroundings to choose a new nest. This process has to work well or the consequences are lethal. 'Organisms live and die by the amount of information they acquire about their environment,' as Andreas Wagner expresses it.[27] Being a good prediction machine means having the ability to learn from experience so as to better anticipate the future and make a smart move. To be efficient, however, a predictive system has to be choosy about what information it stores; it would be wasteful to remember everything. All this requires some sort of internal representation of the world – a type of virtual reality – incorporating sophisticated statistical assessments.[28] Even a bacterium is a wiz at mathematics, it seems.

Summarizing these higher functions, we might say that biological information is not merely acquired, it is *processed*. Shannon's information theory can quantify the number of bits in a cell or an entire organism, but if the name of the game is information processing, then we need to look beyond mere bits and appeal to the theory of *computation*.

Living organisms are not just bags of information: *they are computers*. It follows that a full understanding of life will come only from unravelling its computational mechanisms. And that requires an excursion into the esoteric but fascinating foundations of logic, mathematics and computing.

3

The Logic of Life

'Creativity in biology is not that different from creativity in mathematics.'

– *Gregory Chaitin*[1]

The story of life is really two narratives tightly interwoven. One concerns complex chemistry, a rich and elaborate network of reactions. The other is about information, not merely passively stored in genes but coursing through organisms and permeating biological matter to bestow a unique form of order. Life is thus an amalgam of two restlessly shifting patterns, chemical and informational. These patterns are not independent but are coupled together to form a system of cooperation and coordination that shuffles bits of information in a finely choreographed ballet. Biological information is more than a soup of bits suffusing the material contents of cells and animating it; that would amount to little more than vitalism. Rather, the patterns of information control and organize the chemical activity in the same manner that a program controls the operation of a computer. Thus, buried inside the ferment of complex chemistry is a web of logical operations. *Biological information is the software of life.* Which suggests that life's astonishing capabilities can be traced right back to the very foundations of logic and computation.

A pivotal event in the history of computation was a lecture delivered by the distinguished German mathematician David Hilbert in 1928 to an international congress in Bologna, Italy. Hilbert used the occasion to outline his favourite unanswered mathematical problems. The most profound of these concerned the internal consistency of the subject itself. At root, mathematics is nothing but an elaborate set of definitions, axioms* and the logical deductions flowing from them. We take it for granted that it works. But can we be absolutely rock-solidly *sure* that all pathways of reasoning proceeding from this austere foundation will *never* result in a contradiction? Or simply fail to produce an answer? You might be wondering, *Who cares?* Why does it matter whether mathematics is consistent or not, so long as it works for practical purposes? Such was the mood in 1928, when the problem was of interest to only a handful of logicians and pure mathematicians. But all that was soon to change in the most dramatic manner.

The issue as Hilbert saw it was that, if mathematics *could* be proved consistent in a watertight manner, then it would be possible to test any given mathematical statement as either true or false by a purely mindless handle-turning procedure, or algorithm. You wouldn't need to *understand* any mathematics to implement the algorithm; it could be carried out by an army of uneducated employees (paid calculators) or a machine, cranking away for as long as it took. Is such an infallible calculating machine possible? Hilbert didn't know, and he dignified the conundrum with the title *Entscheidungsproblem* (in English 'the decision problem', but usually referred to as 'the halting problem'.) The term was chosen to address the basic issue of whether some computations might simply go on for ever: they would *never halt*. The hypothetical machine might grind away for all eternity with no answer forthcoming. Hilbert was not interested in the practical matter of how long it might take to get an answer, only whether the machine would reach the end of the procedure in a finite time and output one of two answers: true or false. It may seem reasonable to expect the answer always to be yes. What could possibly go wrong?

Hilbert's lecture was published in 1929, the same year that Szilárd's demon paper appeared. These two very different thought experiments – a calculating engine that may not halt and a thermodynamic engine that may generate perpetual motion – turn out to be

---

* An axiom is a statement taken to be obviously true – such as 'if $x = y$, then $y = x$', which is to say that if on a farm there are the same number of sheep as goats, then there are the same number of goats as sheep.

intimately connected. At the time, however, neither man was aware of that. Nor did they have an inkling that, deep inside biology's magic puzzle box, concealed by layer upon layer of bewildering complexity, it was the incessant drumbeat of mathematics that bestowed the kiss of life.

TO INFINITY AND BEYOND

Mathematics often springs surprises, and at the time of Hilbert's lecture trouble was already brewing in the logical foundations of the subject.* There had been earlier attempts to prove the consistency of mathematics, but in 1901 they were startlingly derailed by the philosopher Bertrand Russell, who identified a famous paradox that lurks inside all formal systems of reasoning. The essence of Russell's paradox is easily described. Consider the following statement, labelled A:

A: This statement is false.

Suppose we now ask: is A true or false? If A is true, then the statement itself declares A to be false. But if A is false, then it is true. By referring to itself in a contradictory way, A seems to be both true *and* false, or neither. We might say it is undecidable. Because mathematics is founded on logic, after Russell the entire basis of the discipline began to look shaky. Russell's paradoxes of self-reference set a time bomb ticking that was to have the most far-reaching consequences for the modern world.

It took the work of an eccentric and reclusive Austrian logician named Kurt Gödel to render the full import of self-referential paradoxes evident. In 1931 he published a paper demonstrating that no consistent system of axioms exists that can prove *all* true statements of arithmetic. His proof hinged on the corrosive existence of self-referential relationships, which imply that there will always be true arithmetic statements that cannot *ever* be proved true within that

* The origins of the Entscheidungsproblem can be traced back to an earlier address by Hilbert, delivered in 1900 to the International Congress of Mathematicians at the Sorbonne in Paris.

system of axioms. More generally, it followed that no finite system of axioms can be used to prove its own consistency; for example, the rules of arithmetic cannot themselves be used to prove that arithmetic will always yield consistent results.

Gödel shattered the ancient dream that cast-iron logical reasoning would always produce irrefutable truth. His result is arguably the highest product of the human intellect. All other discoveries about the world of physical things or the world of reason tell us something we didn't know before. Gödel's theorem tells us that the world of mathematics embeds inexhaustible novelty; even an unbounded intellect, a god, can never know everything. It is the ultimate statement of open-endedness.

Constructed as it was in the rarefied realm of formal logic, Gödel's theorem had no apparent link with the physical world, let alone the biological world. But only five years later the Cambridge mathematician Alan Turing established a connection between Gödel's result and Hilbert's halting problem, which he published in a paper entitled 'On computable numbers, with an application to the Entscheidungsproblem'.[2] It proved to be the start of something momentous.

Turing is best known for his role in cracking the German Enigma code in the Second World War, working in secret at Bletchley Park in the south of England. His efforts saved countless Allied lives and shortened the war by many months, if not years. But history will judge his 1936 paper to be more significant than his wartime work. To address Hilbert's problem of mindless computation Turing envisaged a calculating machine rather like a typewriter, with a head that could scan a moving tape and write on it. The tape would be of unlimited length and divided into squares on which symbols (e.g. 1, 0) could be printed. As the tape passed through the machine horizontally and each square reached the head, the machine would either erase or write a symbol on it or leave it alone, and then advance the tape either left or right by one square, and repeat the process, over and over, until the machine halted and delivered the answer. Turing proved that a number was computable if and only if it could be the output of such a machine after a finite (but possibly huge) number of steps. The key idea here was that of a *universal* computer: 'a single machine which can be used to compute any computable sequence'.[3] Here in this simple statement is the genesis of the modern computer, a device we now take for granted.*

From the pure mathematical point of view, the import of Turing's paper is a proof that there isn't, and can never be, an algorithm to solve the Entscheidungsproblem – the halting problem. In plain English, there can be no way to know in advance, for general mathematical statements, whether or not Turing's machine will halt and output an answer of true or of false. As a result, there will always be mathematical propositions that are quite simply *undecidable*. Though one may certainly take a given *decidable* proposition (e.g. eleven is a prime number) and prove it to be true or false, no one can prove that a statement is *undecidable*.

Though the ramifications of Turing's imaginary computing machine proved stunning for mathematicians, it was the practical application that soon assumed urgency. With the outbreak of war, Turing was tasked with turning his abstract brainchild into a working device. By 1940 he had designed the world's first programmable electronic computer. Christened Colossus, it was built by Tommy Flowers at the Post Office's telephone exchange at Dollis Hill in London and installed at the top-secret code-breaking establishment in Bletchley Park. Colossus became fully operational in 1943, a decade before IBM built its first commercial machine. The sole purpose of Colossus was to assist in the British code-breaking effort and so it was built and operated under an exceptionally tight security blanket. For political reasons, the culture of secrecy surrounding Bletchley Park persisted well after the end of the war and is part of the reason why Flowers and Turing often do not receive credit for being the first architects of the computer. It also allowed the initiative for the commercialization of computers to pass to the United States, where wartime work in this area was rapidly declassified.

Although it was primarily directed at mathematicians, Turing's work was to have deep implications for biology. The particular logical architecture embodied in living organisms mirrors the axioms of logic itself. Life's defining property of self-replication springs directly

* Charles Babbage, a century earlier, arrived at the same basic concept, but he made no attempt to offer a formal proof of universal computability.

from the paradox-strewn domain of propositional calculus and self-reference that underpins the concept of computation, in turn opening the way to now-familiar properties like simulation and virtual reality. Life's ability to construct an internal representation of the world and itself – to act as an agent, manipulate its environment and harness energy – reflects its foundation in the rules of logic. It is also the logic of life that permits biology to explore a boundless universe of novelty, to create 'forms most wonderful', to use Darwin's memorable phrase.

Given that undecidability is enshrined in the very foundations of mathematics, it will also be a fundamental property of a universe based on mathematical laws. Undecidability guarantees that the mathematical universe will always be unbounded in its creative potential. One of the hallmarks of life is its limitless exuberance: its open-ended variety and complexity. If life represents something truly fundamental and extraordinary, then this quality of unconstrained possibility is surely key. Many of the great scientists of the twentieth century spotted the connection between Turing's ideas and biology. What was needed to cement the link with biology was the transformation of a purely computational process into a physical construction process.

A MACHINE THAT COPIES ITSELF

Across the other side of the Atlantic from Alan Turing, the Hungarian émigré John von Neumann was similarly preoccupied with designing an electronic computer for military application, in his case in connection with the Manhattan Project (the atomic bomb). He used the same basic idea as Turing – a universal programmable machine that could compute anything that is computable. But von Neumann also had an interest in biology. That led him to propose the idea of a *universal constructor* (UC), full details of which had to await the posthumous publication of his book *Theory of Self-reproducing Automata*.[4]

The concept of a UC is easy to understand. Imagine a machine that can be programmed to build objects by selecting components from a pool of materials and assembling them into a functional product.

Today we are very familiar with robot assembly lines doing just that, but von Neumann had in mind something more ambitious. Robotic systems are not UCs: a car assembly line can't build a fridge, for example. To be a truly *universal* constructor, the UC has to be able to build *anything* that is in principle constructible, subject to a supply of components. And now here comes the twist that connects Gödel, Turing and biology. A UC also has to be able to build a copy of *itself*. Remember that it was precisely the paradox of self-reference that led Turing to the idea of a universal computer. The idea of a self-reproducing machine thus opens the same logical can of worms. Significantly, living organisms seem to be actual self-reproducing machines. We thus gain insight into the logical architecture of life by deliberating on the concepts of a universal computer (Turing machine) and a universal constructor (von Neumann machine).

An important point von Neumann stressed is that it is not enough for a UC simply to make a replica of itself. It also has to replicate the *instructions* for how to make a UC and insert those instructions into the freshly minted replica; otherwise, the UC's progeny would be sterile. These days we think of robotic instructions as being invisibly programmed into a computer that drives the robot, but to better see the logic of self-reproducing machines it is helpful to think of the instructions imprinted on a punched tape of the sort that drives an old-fashioned pianola (and close to the concept of the tape in a Turing machine). Imagine that the UC has a punched tape fed into it telling it how to make this or that object and that the machine blindly carries out the instructions imprinted on the tape. Among the many possible tapes, each peppered with strategically located holes, there will be one with a pattern of holes that contains the instructions for building the UC itself. This tape will chug through the machine and the UC will build another UC. But, as stated, that's not enough; the mother UC now has to make a copy of the *instruction tape*. For this purpose, the tape now has to be treated not as a set of instructions but as just another physical object to be copied. In modern parlance, the tape must undergo a change of status from being software (instructions) into being hardware – some material with a certain pattern of holes. Von Neumann envisaged what he called a supervisory unit to effect the switch, that is, to toggle between hardware and software as the circumstances demanded. In the final act of the drama, the blindly copied instruction tape is added to the newly made UC to complete the cycle. The crucial insight von Neumann had is that the information on the tape must be treated *in two distinct ways*. The first is as active *instructions* for the UC to build something. The second is as passive *data* simply to be copied as the tape is replicated.

Life as we know it reflects this dual role of information. DNA is both a physical object and an instruction set, depending on circumstances. When a cell is just getting on with life, and this or that protein is needed for some function, the instructions for building the relevant protein are read out from DNA and the protein is made by a ribosome. In this mode, DNA is acting as software. But when the time comes for the cell to replicate and divide, something quite different happens. Special enzymes come along and blindly copy the DNA (including any accumulated flaws) so there is one copy available for each cell after division takes place.* So the logical organization of a living cell closely mirrors that of a von Neumann self-replicating machine. What is still a mystery, however, is the biological equivalent of the supervisory unit that determines when instructions need to switch to become passive data. There is no obvious component in a cell, no special organelle that serves as 'the strategic planner' to tell the cell how to regard DNA (as software or hardware) moment by moment. The decision to replicate depends on a large number of factors throughout the cell and its environment; it is not localized in one place. It provides an example of what is known as epigenetic control involving top-down causation,⁵ a topic I shall discuss in detail later.

Von Neumann recognized that replication in biology is very different from simple copying. After all, crystals grow by copying. What makes biological replication non-trivial is its ability to evolve. If the copying process is subject to errors, and the errors are also copied,

---

* As usual, things are a bit more complicated, because both read-out and replication can occur at the same time, sometimes in the same region, risking traffic accidents. To minimize the potential chaos, genomes organize things to avoid RNA and DNA polymerases going opposite ways. In von Neumann terms, this means that life makes sure the tape moves in only one direction when the system needs to be both hardware and software at the same time.

then the replication process is evolvable. Heritable errors are, of course, the driver of Darwinian evolution. If a von Neumann machine is to serve as a model for biology, it must incorporate the two key properties of self-replication and evolvability.

The idea of von Neumann machines has penetrated the world of science fiction and spawned a certain amount of scaremongering. Imagine a mad scientist who succeeds in assembling such a device and releasing it into the environment. Given a supply of raw materials, it will just go on replicating and replicating, appropriating what it needs until the supply is exhausted. Dispatched into space, von Neumann machines could ravage the galaxy and beyond. Of course, living cells are really a type of von Neumann machine, and we know that a predator let loose can decimate an ecosystem if it spreads unchecked. Terrestrial biology, however, is full of checks and balances arising from the complex web of life, with its vast number of interdependent yet different types of organism, so the damage from unconstrained multiplication is limited. But a solitary replicating interstellar predator may be a different story altogether.

LIFE AS A GAME

Although von Neumann didn't attempt to build a physical self-reproducing machine, he did devise a clever mathematical model that captures the essential idea. It is known as a cellular automaton (CA), and it's a popular tool for investigating the link between information and life. The best-known example of a CA is called, appropriately enough, the Game of Life, invented by the mathematician John Conway and played on a computer screen. I need to stress that the Game of Life is very far removed from real biology, and the word 'cell' in cellular automata is not intended to have any connection with living cells – that's just an unfortunate terminological coincidence. (A prison cell is a closer analogy.) The reason for studying cellular automata is because, in spite of their tenuous link with biology, they capture something deep about the *logic* of life. Simple it might be, but the Game embeds some amazing and far-reaching properties. Small wonder then that it has something of a cult following; people like playing it, even setting it to music, mathematicians enjoy exploring its arcane properties, and biologists mine it for clues about what makes life tick at the most basic level of its organizational architecture.

This is how the Game works. Take an array of squares, like a chessboard or pixels on a computer screen. Each square may either be filled or not. The filled squares are referred to as 'live', the empty ones as 'dead'. You start out with some pattern of live and dead squares – it can be anything you like. To make something happen there must be rules to change the pattern. Every square has eight neighbouring squares: in a simple CA, how a given square changes its state (live or dead) depends on the state of those eight neighbours. These are the rules Conway chose:

1. Any live cell with fewer than two live neighbours dies, as if caused by underpopulation.
2. Any live cell with two or three live neighbours lives on to the next generation.
3. Any live cell with more than three live neighbours dies, as if by overpopulation.
4. Any dead cell with exactly three live neighbours becomes a live cell, as if by reproduction.

The rules are applied simultaneously to every square in the array and the pattern (generally) changes – it is 'updated'. The rules are applied repeatedly, each step being one 'generation', creating shifting patterns that can have a rather mesmeric effect. The real interest of the game, however, is less for art or amusement and more as a tool for studying complexity and information flow among the shapes. Sometimes the patterns on a computer screen seem to take on a life of their own, moving across the screen coherently, or colliding and creating new shapes from the debris. A popular example is called the glider, a cluster of five filled squares that crawls across the screen with a tumbling motion (see Fig. 8). It is surprising that such compelling complexity can arise from the repeated application of simple rules.*

---

* Which raises a rather deep question about the nature of life. Everyone agrees life is complex, but is its complexity because it is the product of a complex *process*, or might it be the outcome of successive simple processes, as in the Game of Life? I am grateful to Sara Walker for stressing the distinction between a complex process and a complex state.

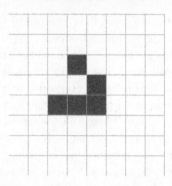

Fig. 8. The Game of Life. This configuration of filled squares, when evolved step by step using Conway's rules, glides across the screen without change of internal configuration. It will continue on its way unless it collides with other filled squares.

Given a random starting pattern, several things can happen in the Game. The patterns may evolve and shift for a while but end up vanishing, leaving the screen blank. Or they may be hung up in static shapes, or cycle through the same shapes again and again every few generations. More interestingly, they may go on for ever, generating unlimited novelty – just like in real biology. How can one know in advance which starting patterns will generate unbounded variety? It turns out that, generally, you can't know. The patterns aren't arbitrary but obey higher-level rules of their own. It has been proved that the patterns themselves can implement basic logical operations in their behaviour. They are a computer inside a computer! The *patterns* can thus represent a Turing machine or a universal computer, albeit a slow one. Because of this property, Turing's undecidability analysis can be directly applied to the Game of Life. Conclusion: one cannot in any systematic way decide in advance whether a given initial pattern settles down or runs on for ever.[6]

I still find it a bit eerie that patterns on a computer screen can become unshackled from their substrate and create a universe of their own, *Matrix*-like, while still being tied to the iron rule of logic in their every move. But such is the power of Gödelian undecidability: the strictures of logic are compatible with the creation of unpredictable novelty. However, the Game of Life does prompt some serious questions about cause and effect. Can we really treat the shapes on the screen as 'things' able to 'cause' events, such as the untidy detritus of collisions? The shapes are, after all, not physical objects but *informational* patterns; everything that happens to them can be explained at the lower level of the computer program. Yet the fundamental undecidability inherent in the system means that there is room for emergent order. Evidently, higher-level informational 'rules of engagement' can be formulated at the level of shapes. Something like this must be going on in life (and consciousness), where the causal narrative can be applied to informational patterns *independently of the physical substrate*.

Though it is tempting to think of the shapes in the Game of Life as 'things' with some sort of independent existence obeying certain rules, there remains a deep question: in what sense can it be said that the collision of two shapes 'causes' the appearance of another? Joseph Lizier and Mikhail Prokopenko at the University of Sydney tried to tease out the difference between mere correlation and physical causation by performing a careful analysis of cellular automata, including the Game of Life.[7] They treated information flowing through a system as analogous to injecting dye into a river and searching for it downstream. Where the dye goes is 'causally affected' by what happens at the injection point. Or, to use a different image, if A has a causal effect on B, it means that (metaphorically speaking) wiggling A makes B wiggle too, a little later. But Lizier and Prokopenko also recognized the existence of what they term 'predictive information transfer', which occurs if simply *knowing something* about A helps you to know better what B might do next, even if there is no direct physical link between A and B.* One might say that the behaviour of A and B are correlated via an information pattern that enjoys its

---

* There is no mystery about correlation without causation. Let me give a simple example. Suppose I email my friend Alice in London the password protecting a secret bank account, which she immediately accesses, and a few moments later I send the same message to Bob in New York, who follows suit. A spy monitoring the account might jump to the erroneous conclusion that Alice had given Bob the password, that is, that the information about the password had passed from London to New York, thus causally linking Alice to Bob. But, in fact, Alice's information and Bob's information are correlated not because one caused the other but because they

own dynamic. The conclusion is that information patterns do form causal units and combine to create a world of emergent activity with its own narrative. Iconoclastic though this statement may seem, we make a similar assumption all the time in daily life. For example, it is well known that as people get older they tend to become more conservative in their tastes and opinions. While this is hardly a law of nature, it is a universal feature of human nature, and we all regard 'human nature' as a thing or property with a real existence, even though we know that human thoughts and actions are ultimately driven by brains that obey the laws of physics.

There are many ways in which CAs can be generalized. For example, Conway's rules are 'local' – they involve only nearest neighbours. But non-local rules, in which a square is updated by reference to, say, the neighbours both one and two squares away, are readily incorporated. So are asynchronous update rules, whereby different squares are updated at different steps. Another generalization is to permit squares to adopt more than two states, rather than simply being 'live' or 'dead'. Von Neumann's main motivation, remember, was to construct a CA that would have the property of both self-reproduction and evolvability. Conway's Game of Life is provably evolvable, but can it also support self-reproduction? Yes, it can. On 18 May 2010 Andrew J. Wade, a Game of Life enthusiast, announced he had found a pattern, dubbed Gemini, that does indeed replicate itself after 34 million generations. On 23 November 2013 another Game of Life devotee, Dave Greene, announced the first replicator that creates a complete copy of itself, including the analogue of the crucial instruction tape, as von Neumann specified. These technical results may seem dry, but it is important to understand that the property of self-replication reflects an extremely special aspect of the Game's logic. It would not be the case for an arbitrary set of automaton rules, however many steps were executed.

All of which brings me to an important and still-unanswered scientific question that flows from von Neumann's work. What is the *minimum* level of complexity needed to attain the twin features of

---

were caused by a common third party (me). Conflating correlation with causation is an easy trap to fall into.

non-trivial replication and open-ended evolvability? If the complexity threshold is quite low, we might expect life to arise easily and be widespread in the cosmos. If it is very high, then life on Earth may be an exception, a freak product of a series of highly improbable events. Certainly the cellular automaton that von Neumann originally proposed was pretty complex, with each square being assigned one of twenty-nine possible states. The Game of Life is much simpler, but it requires major computational resources and still represents a daunting level of complexity. However, these are merely worked-out examples, and the field is still the subject of lively investigation. Nobody yet knows the minimal complexity needed for a CA computer model of a von Neumann machine, still less that for a *physical* UC made of molecules.

Recently, my colleagues Alyssa Adams and Sara Walker introduced a novel twist into the theory of cellular automata. Unlike the Game of Life, which plays out its drama across a two-dimensional array of cells, Adams and Walker used a one-dimensional row of cells. As before, cells may be filled or empty. You start with an arbitrary pattern of filled squares and evolve one step at a time using an update rule – an example is shown in Fig. 9. Time runs downwards in the figure: each horizontal line is the state of the CA at that time step, as derived from the row above by application of the rule. Successive applications generate the pattern. The mathematician Stephen Wolfram did an exhaustive study of one-dimensional CAs: there are 256 possible update rules that take into account the nearest-neighbour squares only. As with the Game of Life, some patterns are boring, for example they become hung up in one state or cycle repeatedly among the same few states. But Wolfram discovered that there are a handful of rules that generate far greater complexity. Fig. 10 shows one example, using Wolfram's Rule 30 and a single filled square as an initial condition. Compare the regularity of Fig. 9 (which uses Rule 90) with the elaborate structure of Fig. 10 (which uses Rule 30).

Adams and Walker wanted a way to make the CA a more realistic representation of biology by including a changing environment, so they coupled two CAs together (computationally speaking): one CA represented the organism, another the environment. Then they introduced a fundamental departure from conventional CA models: they

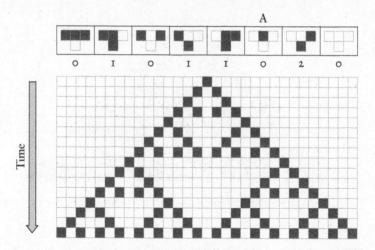

Fig. 9. One-dimensional (elementary) cellular automaton – Wolfram Rule 90. The long box at the top shows the rule structure. Starting with a single filled square in the middle of the first line of the automaton, the pattern below it is generated by applying the rule to each square repeatedly. For example, at the initial step (top row), the single filled square corresponds to the layout in box A (with neighbours either side empty), so that square changes from filled to empty at the next step.

allowed the update rule for the 'organism' to *change*. To determine which of the 256 rules to apply at each step they bundled the 'organism' CA cells into adjacent triplets (that is, 000, 010, 110, and so on) and compared the relative frequencies of each triplet with the same patterns in the 'environment' CA. (If this seems convoluted and technical, don't worry; the details don't matter, just the general idea that introducing non-local rules can be a powerful way to generate novel forms of complexity.) So this arrangement changes the update rule as a function of both the state of the 'organism' itself – making it self-referential – and of the 'environment' – making it an open system. Adams and Walker ran thousands of case studies on a computer to look for interesting patterns. They wanted to identify evolutionary behaviour that is both open-ended (that is, didn't quickly cycle back to the starting state) and innovative. Innovation in this context means

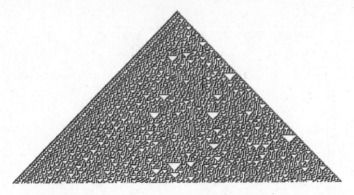

Fig. 10. Rule 30 cellular automaton, showing the evolution of complexity.

that the observed sequence of states could never occur in *any* of the 256 possible fixed-rule CAs, even taking into account every possible starting state. It turned out such behaviour was rare, but there were some nice clear-cut examples. It took a lot of computing time, but they discovered enough to be convinced that, even in this simple model, *state-dependent dynamics* provide novel pathways to complexity and variety.[8] Their work nicely illustrates that merely processing bits of information isn't enough; to capture the full richness of biology, the information-processing rules themselves must evolve. I shall return to this important theme in the Epilogue.

CAN A BIOLOGIST FIX A RADIO?

Whatever the minimal complexity for life may be, there is no doubt that even the simplest known life form is already stupendously complex. Indeed, life's complexity is so daunting that it is tempting to give up trying to understand it in physical terms. A physicist may be able to give an accurate account of a hydrogen atom, or even a water molecule, but what hope is there for describing a bacterium in the same terms?

A generation or two ago things looked a lot brighter. Following the elucidation of the structure of DNA and the cracking of the universal genetic code, biology was gripped by reductionist fervour. There was a tendency to think that answers to most biological questions were to be found at the level of genes, a viewpoint eloquently articulated by Richard Dawkins with his concept of the selfish gene.[9] And there is no doubt that reductionism applied to biology has scored some notable successes. For example, specific defective genes have been linked to a number of heritable conditions such as Tay-Sachs syndrome. But it soon became clear that there is generally no simple connection between a gene, or a set of genes, and a biological trait at the level of the organism. Many traits emerge only when the system as a whole is taken into account, including entire networks of genes in interaction, plus many non-genetic, or so-called epigenetic, factors that may also involve the environment (a topic to which I shall return in the next chapter). And when it comes to social organisms – for example, ants, bees and humans – a complete account requires consideration of the collective organization of the whole community. As these facts sank in, biology began to look hopelessly complex again.

But perhaps all is not lost. The flip side of reductionism is emergence – the recognition that new qualities and principles may emerge at higher levels of complexity that can themselves be relatively simple and grasped without knowing much about the levels below. Emergence has acquired something of a mystical air but in truth it has always played a role in science. An engineer may fully understand the properties of steel girders without the need to consider the complicated crystalline structure of metals. A physicist can study patterns of convection cells knowing nothing about the forces between water molecules. So can 'simplification from emergence' work in biology too?

Confronting this very issue, the Russian biologist Yuri Lazebnik wrote a humorous essay entitled 'Can a biologist fix a radio?'[10] Like radio receivers, cells are set up to detect external signals that trigger appropriate responses. Here's an example: EGF (epidermal growth factor) molecules may be present in tissues and bind to receptor molecules on the surface of a particular cell. The receptor straddles the cell membrane and communicates with other molecules in the cell's innards. The EGF binding event triggers a signalling cascade inside the cell, resulting in altered gene expression and protein production which, in this case, leads to cell proliferation. Lazebnik pointed out that his wife's old transistor radio is also a signal transducer (it turns radio waves into sound) and, with hundreds of components, about as complex as a signal transduction mechanism in a cell.

Lazebnik's wife's radio had gone wrong and needed fixing. How, wondered Lazebnik, might a reductionist biologist tackle the problem? Well, the first step would be to acquire a large number of similar radios and peer into each, noting the differences and cataloguing the components by their colour, shape, size, and so on. Then the biologist might try removing one or two elements or swapping them over to see what happened. Hundreds of learned papers could be published on the results obtained, some of them puzzling or contradictory. Prizes would be awarded, patents granted. Certain components would be established as essential, others less so. Removing the essential parts would cause the radio to stop completely. Other parts might affect only the quality of the sound in complex ways. Because there are dozens of components in a typical transistor radio, linked together in various patterns, the radio would be pronounced 'very complex' and possibly beyond the ability of scientists to understand, given how many variables are involved. Everyone would agree, however, that a much bigger budget would be needed to extend the investigation.

In the expanded research programme a useful line of inquiry would be to use powerful microscopes to look for clues *inside* the transistors and capacitors and other objects, right down to the atomic level. The huge study might well go on for decades and cost a fortune. And it would, of course, be useless. Yet what Lazenbik describes in the transistor radio satire is precisely the approach of much of modern biology. The major point that the author wanted to make is that an *electronic engineer*, or even a trained technician, would have little difficulty fixing the defective radio, for the simple reason that this person would be well versed in *the principles of electronic circuitry*. In other words, by understanding how radios work and how the parts are wired together to achieve well-defined functions, the task of fixing a defective model is rendered straightforward. A few carefully chosen tweaks, and the music plays again. Lazenbik laments that biology has not attained this level of understanding and that few biologists even think about life in those terms – in terms of living cells containing modules which have certain logical functions and are 'wired together', chemically speaking, to form networks with feedback, feed-forward, amplification, transduction and other control functions to attain collective functionality. The main point is that in most cases it is *not necessary* to know what is going on *inside* those modules to understand what is happening to the system as a whole.

Fortunately, times are changing. The very notion of life is being reconceptualized, in a manner that closely parallels the realms of electronics and computing. A visionary manifesto for a future systems biology along these lines was published in *Nature* in 2008 by the Nobel prizewinning biologist Paul Nurse, soon to become President of The Royal Society. In a paper entitled 'Life, logic and information', Nurse heralded a new era of biology.[11] Increasingly, he pointed out, scientists will seek to map molecular and biochemical processes into the biological equivalent of electronic circuit boards:

> Focusing on information flow will help us to understand better how cells and organisms work ... We need to describe the molecular interactions and biochemical transformations that take place in living organisms, and then translate these descriptions into the logic circuits that reveal how information is managed ... Two phases of work are required for such a programme: to describe and catalogue the logic circuits that manage information in cells, and to simplify analysis of cellular biochemistry so that it can be linked to the logic circuits ... A useful analogy is an electronic circuit. Representations of such circuits use symbols to define the nature and function of the electronic components used. They also describe the logic relationships between the components, making it clear how information flows through the circuit. A similar conceptualization is required of the logic modules that make up the circuits that manage information in cells.

Philosophers and scientists continue to bicker over whether, 'in principle', all biological phenomena could be reduced solely to the goings-on of atoms, but there is agreement that, as a practical matter, it makes far more sense to search for explanations at higher levels. In electronics, a device can be perfectly well designed and assembled from standard components – transistors, capacitors, transformers, wires, and so on – without the designer having to worry about the precise processes taking place in each component at the atomic level. You don't have to know how a component *works*, only what it *does*. And where this practical approach becomes especially powerful is when the electronic circuit is processing information in some way – in signal transduction, rectification or amplification, or as a component in a computer – because then the explanatory narrative can be cast entirely in terms of information flow and software, without any reference back to the hardware or module itself, still less its molecular parts. In the same vein, urges Nurse, we should seek, where possible, explanations for processes within a cell, and between cells, based on the informational properties of the higher-level units.

When we look at living things we see their material bodies. If we probe inside, we encounter organs, cells, sub-cellular organelles, chromosomes and even (with fancy equipment) molecules themselves. What we don't see is information. We don't see the swirling patterns of information flows in the brain's circuitry. We don't see the army of demonic information engines in cells, or the organized cascades of signalling molecules executing their restless dance. And we don't see the densely packed information stored in DNA. What we see is stuff, not bits. We are getting only half the story of life. If we could view the world through 'information eyes', the turbulent, shimmering information patterns that characterize life would leap out as distinctive and bizarre. I can imagine an artificial intelligence (AI) of the future being tuned to information and being trained to recognize people not from their faces but from the informational architecture in their heads. Each person might have their own identification pattern, like the auras of pseudoscience. Importantly, the information patterns in living things are not random. Rather, they have been sculpted by evolution for optimal fitness, just as have anatomy and physiology.

Of course, humans cannot directly perceive information, only the material structures in which it is instantiated, the networks in which it flows, the chemical circuitry that links it all together. But that does not diminish the importance of information. Imagine if we tried to understand how a computer works by studying only the electronics inside it. We could look at the microchip under a microscope, study the wiring diagram in detail and investigate the power source. But we would still have no idea how, for example, Windows performs its magic. To fully understand what appears on your computer screen you have to consult a software engineer, one who writes computer code to create the functionality, the code that organizes the bits of information whizzing around the circuitry. Likewise, to fully explain life we need to understand *both* its hardware *and* its software – its molecular organization *and* its informational organization.

BIOLOGICAL CIRCUITS AND THE MUSIC OF LIFE

Mapping life's circuitry is a field still in its infancy and forms part of the subject known as systems biology. Electronic circuits have components that are well understood by physicists. The biological equivalent is not so well understood. Many chemical circuits are controlled by genes 'wired' together via chemical pathways to create features like feedback and feed-forward – familiar from engineering, but the details can be messy. To give the flavour, let me focus on a very basic property of life: regulating the production of proteins. Organisms cleverly monitor their environment and respond appropriately. Even bacteria can detect changes around them, process that information and implement the necessary instructions to alter their state to advantage. Mostly, the alteration involves boosting or suppressing the production of certain proteins. Making the right amount of a particular protein is a delicately balanced affair that needs to be carefully tuned. Too much could be toxic; too little may mean starvation. How does a cell regulate *how much* of a particular protein is needed at any given time? The answer lies with a set of molecules (themselves proteins) known as transcription factors, with distinctive shapes that recognize specific segments of DNA and stick to them. Thus bound, they serve to increase or decrease the rate at which a nearby gene is expressed.

It's worth understanding precisely how they do this. Earlier (p. 61), I discussed a molecule called RNA polymerase whose job it is to crawl along DNA and 'read out' the sequence, creating a matching molecule of RNA as it goes. But RNA polymerase doesn't just do this whimsically. It waits for a signal. ('My protein is needed: transcribe me now!') There is a region of DNA near the start of the gene that issues the 'go' signal; it's called a promoter, because it promotes the transcription process. The RNA is attracted to the promoter and will bind to it to initiate transcription: docking followed by chugging, colloquially speaking. But RNA will dock to the promoter only if the latter is in 'go' mode. And it is here that transcription factors regulate what happens. By binding to the promoter region, a given transcription factor can block it and frustrate the RNA docking manoeuvre. In this role, the transcription factor is known, for obvious reasons, as a repressor. All this is fine if the protein isn't needed. But what happens if circumstances change and the blocked gene needs to be expressed? Obviously, the blocking repressor molecule has to be evicted somehow. Well, how does *that* step work?

A good example, figured out long ago, is a mechanism used by the commonplace bacterium *E. coli*. Glucose tops the bacterium's favourite menu, but if glucose is in short supply this versatile microbe can muddle through by metabolizing another sugar called lactose. To accomplish the switch, the bacterium needs three special proteins, requiring three adjacent genes to be expressed. It would be wasteful to keep these genes active just as a contingency plan, so *E. coli* has a chemical circuit to regulate the on–off function of the requisite genes. When glucose is plentiful, a repressor transcription factor binds to the promoter region of DNA, close to the three genes, and blocks RNA polymerase from binding and beginning the transcription process of the said genes: the genes remain off. When glucose is unavailable but there is lactose around, a by-product of the lactose binds to the repressor molecule and inactivates it, opening the way for RNA polymerase to attach to the DNA and do its stuff. The three key genes are then expressed and lactose metabolism begins. There is another switching mechanism to turn the lactose genes off again when glucose becomes plentiful once more.

In all, *E. coli* has about 300 transcription factors to regulate the production of its 4,000 proteins. I have described a repressor function, but other chemical arrangements permit other transcription factors to serve as activators. In some cases, the same transcription factor can activate many genes, or activate some and repress others.

These various alternatives can lead to a rich variety of functions.[12] (For comparison, humans have about 1,400 transcription factors for their 20,000 genes.)[13]

Box 6: How cells do computation

Transcription factors may combine their activities to create various logic functions similar to those used in electronics and computing. Consider, for example, the AND function, where a switch Z is turned on only if a signal is received from switches X *and* Y together. To implement this, a chemical signal flips the transcription factor X into its active shape X*; X is switched on, chemically speaking. Thus activated, X* may then bind to the promoter of gene Y, causing Y to be produced. If, now, there is a second (different) signal that switches Y to *its* active form, Y*, the cell has both X* and Y* available together. This arrangement can serve as an AND logic gate if there is a third gene, Z, designed (by evolution!) to be switched on only if X* and Y* are present together and bind to its promoter. Other arrangements can implement the OR logic operation, whereby Z is activated if *either* X *or* Y is converted to its active form and binds to Z's promoter. When sequences of such chemical processes are strung together, they can form circuits that implement cascades of signalling and information processing of great complexity. Because transcription factors are themselves proteins produced by other genes regulated by other transcription factors, the whole assemblage forms an information-processing and control network with feedback and feed-forward functions closely analogous to a large electronic circuit. These circuits facilitate, control and regulate patterns of information in the cell.

Given the vast number of possible combinations of molecular components and chemical circuits, you might imagine that the information flow in a cell would be an incomprehensible madhouse of swirling bits. Remarkably, it is far more ordered. There are many recurring themes, or informational motifs, across a wide range of networks, suggesting a high utility for certain biological functions. One example is the feed-forward loop, the basic idea of which I introduced above in connection with *E. coli* metabolism. Taking into account the possibility that the logic functions can be either AND or OR gates (see Box 6), there are thirteen possible gene regulation combinations, and of those only one, the feed-forward loop, is a network motif.[14] Since it is rather easy for a mutation in a gene to remove a link in a chemical network, the fact that certain network motifs survive so well suggests strong selection pressure at work in evolution. There must be a reason why these recurring motifs are, literally, vital. One explanation is robustness. Experience from engineering indicates that when the environment is changing a modular structure with a small range of components adapts more readily. Another explanation is versatility. With a modest-sized toolkit of well-tried and reliable parts, a large number of structures can be built using the same simple design principles in a hierarchical manner (as Lego enthusiasts and electronic engineers know well).[15]

Although I have focused on transcription factors, there are many other complex networks involved in cellular function, such as metabolic networks that control the energetics of cells, signal transduction networks involving protein–protein interactions, and, for complex animals, neural networks. These various networks are not independent but couple to each other to form nested and interlocking information flows. There are also many additional mechanisms for transcription factors to regulate cellular processes, either individually or in groups, including acting on mRNA directly or modifying other proteins in a large variety of ways. The existence of so many regulatory chemical pathways enables them to fine-tune their behaviour to play 'the music of life' by responding to external changes with a high degree of fidelity, much as a well-tuned transistor radio can flawlessly play the music of Beethoven.

In more complex organisms, gene control is likewise more complex. Eukaryotic cells, which have nuclei, package most of their DNA into several chromosomes (humans have twenty-three). Within the chromosomes, DNA is tightly compacted, wrapped around protein spindles and further folded and squashed up to a very high degree. In this compacted form the material is referred to as chromatin. How the chromatin is distributed within the nucleus depends on a number of factors, such as where the cell might be in the cell cycle. For much of the cycle the chromatin remains tightly bound, preventing the genes being 'read' (transcribed). If a protein coded by a gene or a set of genes is needed, the architecture of the chromatin has to change to enable the read-out machinery to gain access to the requisite segments. Reorganization of chromatin is under the control of a network of threads, or microtubules. Thus, whole sets of genes may be silenced or activated *mechanically*, either by keeping them 'under wraps' (wrapped up, more accurately) or by unravelling the highly compacted chromatin in that region of the chromosome to enable transcription to proceed. There is more than a faint echo here of Maxwell's demon. In this case, the nuclear demons quite literally 'pull the strings' and open, not a shutter, but an elaborately wound package that encases the relevant information-bearing genes. Significantly, cancer cells often manifest a noticeably different chromatin architecture, implying an altered gene expression profile; I shall revisit that topic in the next chapter.

As scientists unravel the circuit diagrams of cells, many practical possibilities are opening up that involve 'rewiring'. Bio-engineers are busy designing, adapting, building and repurposing living circuitry to carry out designated biological tasks, from producing new therapeutics to novel biotech processes – even to perform arithmetic. This 'synthetic biology' is mostly restricted to bacteria but, recently, new technology has enabled this type of work to be extended to mammalian cells. A technique called Boolean logic and arithmetic through DNA excision, or BLADE, has been developed at Boston University and the ETH in Zurich.[16] The researchers can build quite complex logic circuits to order, and foresee being able to use them to control gene expression. Many of the circuits they have built seem to be entirely new, that is, they have never been found in existing organisms. The group of Hideki Koboyashi at Boston University finds the promise of rewiring known organisms compelling: 'Our work represents a step toward the development of "plug-and-play" genetic circuitry that can be used to create cells with programmable behaviors.'[17] Currently, synthetic circuits are a rapidly expanding area of research in systems biology with more publications detailing novel circuits published every year.[18] The medical promise of this new 'electronics' approach to life is immense. Where disease (for example, cancer) is linked to defective information management – such as a malfunctioning module or a broken circuit link – a remedy might be to chemically rewire the cells rather than destroy them.*

Box 7: By their bits ye shall know them: the advent of the digital doctor

Imagine a physician of the future (doubtless an AI) who, through some amazing technology that can detect gene expression in real time, would gaze at the dancing, twinkling patterns like city lights seen from afar and diagnose a patient's illness. This would be a digital doctor who deals in bits, not tissues, a medical software engineer. I can imagine my futuristic physician proclaiming that there are early signs of cancer in this or that shimmering cluster, or that an inherited genetic defect is producing an anomalous luminous patch, indicating overexpression of such and such a protein in the liver, or maybe quieter spots suggesting that some cells are not getting enough oxygen, oestrogen or calcium. The study of information flow and information clustering would provide a diagnostic tool far more powerful than the battery of chemical tests used today. Treatment would focus on establishing healthy, balanced information patterns, perhaps by attending to, or even re-engineering, some defective modules, much as an electronics engineer (of old) might replace a transistor or a resistor to restore a radio to proper functionality. (In this respect, what I am describing is reminiscent of some Eastern approaches to medicine.) The digital doctor might not seek to replace any hardware modules but instead decide to rewrite some code and upload it into the patient somehow, at the cellular level, to restore normal functionality – a sort of cellular reboot.

It may seem like science fiction, but information biology is paralleling computer technology, albeit a few decades behind. The 'machine

---

* There is an inevitable downside to all this, which is the possibility of bio-hacking for reasons of control or genocide.

code' for life was cracked in the 1950s and 1960s with the elucidation of the DNA triplet code and the translation machinery. Now we need to figure out the 'higher level' computer language of life. This is an essential next step. Today's software engineers wouldn't design a new computer game by writing down vast numbers of 1s and 0s; they use a higher-level language like Python. By analogy, when a cell regulates, for example, the electric potential across its membrane by increasing the number of protons it pumps out, a 'machine code' description in terms of gene codons isn't very illuminating. The cell as a unit operates at a much higher level to manage its physical and informational states, deploying complex control mechanisms. These regulatory processes are not arbitrary but obey their own rules, as do the higher-level computer languages used by software engineers. And, just as software engineers are able to re-program advanced code, so will bio-engineers redesign the more sophisticated features of living systems.

GENE NETWORKS AS MODULES

Biological circuitry can generate an exponentially large variety of form and function but, fortunately for science, there are some simple underlying principles at work. Earlier in this chapter I described the Game of Life, in which a few simple rules executed repeatedly can generate a surprising degree of complexity. Recall that the game treats squares, or pixels, as simply on or off (filled or blank) and the update rules are given in terms of the state of the nearest neighbours. The theory of networks is closely analogous. An electrical network, for example, consists of a collection of switches with wires connecting them. Switches can be on or off, and simple rules determine whether a given switch is flipped, according to the signals coming down the wires from the neighbouring switches. The whole network, which is easy to model on a computer, can be put in a specific starting state and then updated step by step, just like a cellular automaton. The ensuing patterns of activity depend both on the wiring diagram (the topology of the network) and the starting state. The theory of networks can be developed quite generally as a mathematical exercise: the switches are called 'nodes' and the wires are called 'edges'. From very simple network rules, rich and complex activity can follow.

Network theory has been applied to a wide range of topics in economics, sociology, urban planning and engineering, and across all the sciences, from magnetic materials to brains. Here I want to consider network theory applied to the regulation of gene expression – whether they are switched on or off. As with cellular automata, networks can exhibit a variety of behaviours; the one I want to focus on is when the system settles into a cycle. Cycles are familiar from electronics. For example, there's a new top-of-the-range dishwasher in my kitchen, which I installed myself. Inside it has an electronic circuit board (actually just a chip these days) to control the cycle. There are eight different possible cycles. The electronics has a device to halt the cycle if there is a problem. In that, dishwashers are not alone: the cells in your body have a similar circuit to control their cycles.

What is the cell cycle? Imagine a newborn bacterium – that is to say, the parent bacterium has recently split in two. A daughter cell is just starting out on an independent life. The young bacterium gets busy doing what bacteria have to do, which in many cases involves a lot of just hanging out. But its biological clock is ticking; it feels the need to reproduce. Internal changes take place, culminating in the replication of DNA and fission of the entire cell. The cycle is now complete.*

In complex eukaryotic organisms the cell cycle is more complicated, as you would expect. A good compromise is yeast, which, like humans, is a eukaryote, but it is single-celled. The cell cycle of yeast has received a lot of attention (and a Nobel Prize, shared by Paul Nurse and my ASU colleague Lee Hartwell) and the control circuit that runs the cycle was worked out by Maria Davidich and Stefan Bornholdt at the University of Bremen.[19] In fact, there are many types of yeast. I shall discuss just one, *Schizosaccharomyces pombe*, otherwise known as fission yeast. The relevant network is shown in Fig. 11. The nodes – the blobs

---

* The issue of the biological clock is a fascinating separate story. As I have mentioned, bacteria going flat out can go through the whole cycle in twenty minutes, but hardy microbes living at sub-zero temperatures, called psychrophiles, may take hundreds of years.

in the figure – represent genes (or, strictly, the proteins the genes encode); the edges are the chemical pathways linking genes (analogous to the wires in electronics); the arrows indicate that one gene activates the other; and the barred line indicates that a gene inhibits or suppresses the other (similar to the way that the neighbouring squares in the Game of Life may prompt or inhibit the square being filled or vacated). Notice there are some genes with loopy broken arrows, indicating self-inhibition. Each gene adds up all the pluses ('activate!') and minuses ('suppress!') of the incoming arrows and switches itself on, or off, or stays as it is, according to a specific voting rule.

The job of this network is to take the cell step by step through the cycle, halting the proceedings if something goes wrong and returning

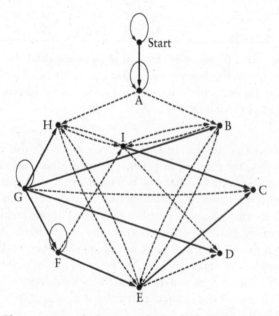

Fig. 11. The gene network that controls the cell cycle of yeast. The nodes represent proteins which may be expressed (1) or not (0) by the associated gene. A solid line indicates that expression of that protein enhances the expression of the other protein represented at the far node; a broken line indicates inhibition.

Table 2

Time step	Start	A	B	C	D	E	F	G	H	I	Cell phase
1	1	0	1	0	1	0	0	0	1	0	Start
2	0	1	1	0	1	0	0	0	1	0	G1
3	0	0	0	0	1	0	0	0	0	0	G1/S
4	0	0	0	0	1	0	0	0	0	1	G2
5	0	0	0	1	0	0	0	0	0	1	G2
6	0	0	0	1	0	1	0	0	0	1	G2/M
7	0	0	0	1	0	1	1	0	0	1	G2/M
8	0	0	0	1	0	0	1	1	0	0	M
9	0	0	1	0	1	0	0	1	1	0	M
10	0	0	1	0	1	0	0	0	1	0	G1

The table represents the step-by-step state of the gene network that controls the cell cycle of yeast, shown in Fig. 11. The letters correspond to the labels assigned to the nodes in the figure; 0 indicates that the relevant node is switched off, 1 that it is on.

the system to its initial state when the cycle is over. In this essential functionality, the network may simply be treated as a collection of interconnected switches that can be modelled on a computer. The gene regulatory network controlling the cell cycle of fission yeast is particularly easy to study because, to a good approximation, the genes involved may be considered either fully on or fully off, not dithering in between. This makes for a pleasing simplification because, mathematically, we may represent 'on' by 1 and 'off' by 0, then make up a rule table with 0s and 1s to describe what happens when the network is started out in some particular state and allowed to run through its little repertoire.*

* Simple networks like this are referred to as Boolean, after George Boole, who introduced the idea of an algebra of logical operations in the nineteenth century.

Using the way I have labelled the genes in Fig. 11, the starting state of the network is A: off; B: on; C: off; D: on; E: off; F: off; G: off; H: on I: off: in binary, that is 010100010. The cycle begins when the node labelled 'start', which sets off the show, flips on (representing an external chemical prompt along the lines of 'Well, go on then, get on with it!'). It is then straightforward to run a computer model of the network step by step and compare the output with reality. Table 2 shows the state of the network at each step. The intermediate states of os and 1s correspond to recognizable physical states which the cell passes through in the cycle. Those physical states are labelled along the right-hand column; the letters stand for biological terms (for example, M stands for 'mitosis'). After ten steps the network returns to the starting state, awaiting a new 'start' signal to initiate the next cycle.

You could make a movie of Fig. 11 in which the nodes light up when they are on and blink out when they are off. There would be a pretty pattern of twinkling lights for ten steps, like fireflies out of kilter. It could perhaps be set to music – the music of life! Let's scale up this fanciful image and envision a human being as a dazzling constellation of twinkling gene-lights forming myriads of swirling patterns – a veritable cacophony if set to music. The star field would be far more than just the genes that control the cycles of all the different cell types. There would be 20,000 genes, all performing their own act. Some lights might stay mostly off, some mostly on, while others would flip on and off in a variety of beats.

The point I want to make is that these shifting information patterns are not random; they portray the organized activity of the network and therefore the organism. And the question is, what can we learn from studying them, from using mathematics and computer models to characterize the shimmering patterns, to track the flow of information, to see where it is stored and for how long, to identify the 'manager genes' and the drones? In short, to build up an

---

They are ideal for applying Shannon's measure of information, and elaborations thereof, based on binary (0 and 1) choices (see p. 36).

informational narrative that captures the essence of what the organism is about, including its management structure, command-and-control architecture and potential failure points.

Well, we can start with yeast: there are only ten nodes and twenty-seven edges in the *Schizosaccharomyces pombe* cell cycle network. Yet even that requires a lot of computing power to analyse. First order of business is to confirm that the patterns are non-random. More precisely, if you just made up a network randomly with the same number of nodes and edges, would the twinkling lights differ in any distinctive way from Mother Nature's yeast network? To answer that, my ASU colleagues Hyunju Kim and Sara Walker ran an exhaustive computer study in which they traced the ebb and flow of information as it swirls around the yeast network.[20] This sounds easy, but it isn't. You can't follow it by eye: there has to be a precise mathematical definition of information transfer (see Box 8). The upshot of their analysis is that there is an elevated and systematic flow of information around the yeast network well in excess of random. Evolution has, it seems, sculpted the network architecture in part for its information-processing qualities.

Box 8: Tracking information flow in gene networks

One may ask of a given network node, say A, whether knowing its history helps in predicting what it will do at the next step. That is, if you look at, say, the three preceding steps of node A and note 'on' or 'off', does that three-step history improve the odds of you correctly guessing on or off for the next step? If it does, then we can say that some information has been *stored* in node A. One can then look at another node, say B, and ask, does knowing the current state of B *improve* the odds of correctly guessing what A will do next, over and above just knowing the history of A? If the answer is yes, it implies that some information has been transferred from B to A. Using that definition, known as 'transfer entropy', my colleagues ranked all pairs of nodes in the yeast cell cycle network by the amount of information transferred, and then compared this rank order with those from an average taken over a thousand random networks. There was a big difference. In a nutshell, the yeast gene network transfers markedly more information than a random one. Digging a little deeper to pin down precisely what is making the difference, the researchers zeroed in on a set of four nodes (B, C, D and H in Fig. 11) that seemed to be calling the shots. The special role of these four genes has earned them the name 'the control kernel'. The control kernel seems to act like a choreographer for the rest of the network, so if one of the other nodes makes a mistake (is on when it should be off, or vice versa), then the control kernel pulls it back into line. It basically steers the whole network to its designated destination and, in biological terms, makes sure the cell fissions on cue with everything in good order. Control kernels seem to be a general feature of biological networks. So in spite of the great complexity of behaviour, a network's dynamics can often be understood by looking at a relatively small subset of nodes.

It would be wrong of me to give the impression that information flow in biology is restricted to gene regulatory networks. Unfortunately, the additional complexity of some other networks makes them even harder to model computationally, especially as the simple version of os and 1s (off and on) mostly won't do. On top of that, the number of components skyrockets when it comes to more finely tuned functions like metabolism. The general point remains: biology will 'stand out' from random complexity in the manner of its information patterning and processing, and though complex, the software account of life will still be vastly simpler than the underlying molecular systems that support it, as it is for electronic circuits.

Network theory confirms the view that information can take on 'a life of its own'. In the yeast network my colleagues found that 40 per cent of node pairs that are correlated via information transfer are not in fact physically connected; there is no direct chemical interaction. Conversely, about 35 per cent of node pairs transfer no information between them even though they *are* causally connected via a 'chemical wire' (edge). Patterns of information traversing the system may appear to be flowing down the 'wires' (along the edges of the graph) even when they are not. For some reason, 'correlation without causation' seems to be amplified in the biological case relative to random networks.

In *Surely You're Joking, Mr. Feynman!*,[21] the raconteur physicist and self-confessed rascal describes how, as a youngster, he gained a reputation for being able to mend malfunctioning radios (yes, that again!). On one occasion he was initially chided for briefly peering into the radio then merely walking back and forth. Being Richard Superbrain Feynman, he had soon figured out the fault and effected a simple repair. 'He fixes radios by thinking!' gushed his dazzled client. The truth is, you generally can't tell just by looking at the layout of an electronic circuit what the problem might be. The performance of a radio depends *both* on the circuit topology and on the physical characteristics of the components. If a resistor, say, is too large or a capacitor too small, the information flow may not be optimal – the output may be distorted. The same is true of all networks – biological, ecological, social or technological. Similar-looking networks can exhibit very different patterns of information flow because their components – the nodes – may have different properties. In the case of the yeast cell cycle, a simple on-or-off rule was used (with impressive results), but there are many different candidate mathematical relationships that could be employed, and they will yield different flow patterns. The bottom line is, there is no obvious relationship between the information pattern's dynamics and the 'circuit' topology. Therefore, for many practical purposes, it pays to treat the information patterns as 'the thing of interest' and forget about the underlying network that supports it. Only if something goes wrong is it necessary to worry about the actual 'wiring'.

Two Israeli mathematicians, Uzi Harush and Baruch Barzel, recently did a systematic study using a computer model of information flow in a broad range of networks. They painstakingly tracked the contribution that each node and pathway made to the flow of information in an attempt to identify the main information highways. To accomplish this, they tried meddling with the system, for example 'freezing' nodes to see how the information flow changed then assessing the difference it might make to the strength of a signal in a specific downstream node. There were some surprises: they found that in some networks information flowed mainly through the hubs (a hub is where many links concentrate, for example, servers in the internet), while in others the information shunned the hubs, preferring to flit around the periphery of the network. In spite of the diversity of results, the mathematicians report that 'the patterns of information flow are governed by universal laws that can be directly linked to the system's microscopic dynamics'.[22] Universal laws? This claim goes right to the heart of the matter of when it is legitimate to think of information patterns as coherent things with an independent existence. It seems to me that if the patterns themselves obey certain rules or laws, then they may be treated as entities in their own right.

COLLECTIVELY SMART

'Go to the ant, thou sluggard; consider her ways, and be wise.'
– Proverbs 6:6

Network theory has found a fruitful application in the subject of social insects, which also display complex organized behaviour deriving from the repeated application of simple rules between neighbouring individuals. I was once sitting on a beach in Malaysia beneath a straw umbrella fixed atop a stout wooden post. I recall drinking beer and eating potato crisps. One of the crisps ended up on the ground, where I left it. A little later I noticed a cluster of small ants swarming around the abandoned object, taking a lot of interest. Before long they set about transporting it, first horizontally across the sand, then vertically up the wooden post. This was a heroic collective effort – it was a big crisp and they were little ants. (I had no idea that ants liked crisps anyway.) But they proved equal to the task. Organized round the periphery, the gals (worker ants are all female) on top pulled, while those underneath pushed. Where were they headed? I noticed a vertical slot at the top of the post with a few ants standing guard.

This must be their nest. But all that pushing and pulling was surely futile because a) the crisp looked too big to fit in the slot and b) the ants would have to rotate the crisp (which was approximately flat) through two right angles to insert it. It would need to project out perpendicular to the post in a vertical plane before the manoeuvre could be executed. Minutes later I marvelled that the ants' strategy had been successful: the crisp was dragged into the slot in one piece. Somehow, the tiny, pin-brained creatures had assessed the dimensions and flatness of the crisp when it lay on the ground and figured out how to rotate it into the plane of the slot. And they did it on the first try!

Stories like this abound. Entymologists enjoy setting challenges and puzzles for ants, trying to outsmart them with little tricks. Food and nice accommodation seem to be their main preoccupations (the ants', that is, though no doubt also the entymologists'), so to that end they spend a lot of time foraging, milling around seemingly at random and seeking out a better place to build a nest. There is a big social-insect research group at ASU run by Stephen Pratt, and a visit to the ant lab is always an entertaining experience. Since almost all ants of the same species look the same, the wily researchers paint them with coloured dots so that they can track them, see what they get up to. The ants don't seem to mind. Although at first glance the scurrying insects look to be taking random paths, they are mostly not. They identify trails based on the shortest distance from the nest and mark them chemically. If, as part of the experiment, their strategy is disrupted by the entymologist, for example by moving a source of food, the ants default to a Plan B while they reassess the local topography. The most distinctive feature of their behaviour is that they communicate with each other. When one ant encounters another, a little ritual takes place that serves to transfer some positional information to the other ant.* In this manner, data gathered by a solitary ant can quickly become disseminated among many in the colony. The way now lies open for collective decision-making.

---

* Another common communication strategy for ants is stigmergy – communication via the environment by depositing and sensing pheromones.

In the case of the purloined crisp, it was clear that no one ant had a worked-out strategy in advance. There was no foreman (forewoman, really) of the gang. The decision-making was done collectively. But how? If I meet a friend on the way home from work and he asks, 'How's yer day goin', mate?' he risks being subjected to five minutes of mostly uninteresting banter (which nevertheless might convey a lot of information). Unless ants are very fast talkers, their momentary encounters would not amount to more than a few logical statements along the lines of 'if, then'. But integrate many ant-to-ant encounters across a whole colony and the power of the collective information processing escalates.

Ants are not alone in their ability to deploy some form of swarm decision-making, even, one might hazard, swarm intelligence. Bird flocks and fish schools also act in unison, swooping and swerving as if all are of one mind. The best guess as to what lies behind this is that the application of some simple rules repeated lots of times can add up to something pretty sophisticated. My ant colleagues at ASU are investigating the concept of 'distributed computation', applying information theory to the species *Temnothorax rugatulus*, which forms colonies with relatively few workers (less than 300), making them easier to track. The goal is to trace how information flows around the colony, how it is stored and how it propagates during nest-building. All this is being done in the lab under controlled conditions. The ants are offered a variety of new nests (the old one is disrupted to give them some incentive to move house), and the investigators study how a choice is made collectively. When ants move en masse, a handful who know the way go back to the nest and lead others along the path: this is called 'tandem running'. It's slow going, as the naïve ants bumble along, continually touching the leaders to make sure they don't get lost (ants can't see very far). When enough ants have learned the landmarks, tandem running is abandoned in favour of piggy-backing, which is quicker.

One thing my colleagues are focusing on is reverse tandem running, where an ant in the know leads another ant from the new nest back to the old one. Why do that? It seems to have something to do with the dynamics of negative feedback and information erasure, but the issue isn't settled. To help things along, the researchers have designed a dummy ant made of plastic with a magnet inside. It is guided by a small robot concealed beneath a board on which the ants move. Armed with steerable artificial ants, my colleagues can create their own tandem runs to test various theories. The entire action is recorded on video for later quantitative analysis. (You can tell that this research is a lot of fun!)

Social insects represent a fascinating middle stage in the organization of life, and their manner of information processing is of special interest. But the vast and complex web of life on Earth is woven from information exchange between individuals and groups at all levels, from bacteria to human society. Even viruses can be viewed as mobile information packets swarming across the planet. Viewing entire ecosystems as networks of information flow and storage raises some important questions. For example, do the characteristics of information flow follow any scaling laws* as you go up in the hierarchy of complexity, from gene regulatory networks through deep-ocean volcanic-vent ecosystems to rainforests? It seems very likely that life on Earth as a whole can be characterized by certain definite information signatures or motifs. If there is nothing special about terrestrial life, then we can expect life on other worlds to follow the same scaling laws and display the same properties, which will greatly assist in the search for definitive bio-signatures on extra-solar planets.

THE MYSTERY OF MORPHOGENESIS

Of all the astonishing capabilities of life, morphogenesis – the development of form – is one of the most striking. Somehow, information etched into the one-dimensional structure of DNA and compacted into a volume one-billionth that of a pea unleashes a choreography of exquisite precision and complexity manifested in three-dimensional space, up to and including the dimensions of an entire fully formed baby. How is this possible?

---

* A scaling law is a mathematical relationship describing how a quantity increases or decreases with scale. For example, in the solar system there are a much greater number of smaller objects like asteroids and moons than there are planets.

In Chapter 1, I mentioned how the nineteenth-century embryologist Hans Dreisch was convinced some sort of life force was at work in embryo development. This rather vague vitalism was replaced by the more precise concept of 'morphogenetic fields'. By the end of the nineteenth century physicists had enjoyed great success using the field concept, originally due to Michael Faraday. The most familiar example is electricity: a charge located at a point in space creates an electric field that extends into the three-dimensional region around it. Magnetic fields are also commonplace. It is no surprise, therefore, that biologists sought to model morphogenesis along similar lines. The trouble was, nobody could give a convincing answer to the obvious question: a field of *what*? Not obviously electric or magnetic; certainly not gravitational or nuclear. So it had to be a type of 'chemical field' (by which I mean chemicals of some sort spread out across the organism in varying concentrations), but the identity of the chemical 'morphogens' long remained obscure.

It was to be many more decades before significant progress was made. In the latter part of the twentieth century biologists began approaching morphogenesis from a genetic standpoint. The story they concocted goes something like this. When an embryo develops from a fertilized egg, the original single cell (zygote) starts out with almost all its genes switched on. As it divides again and again various genes are silenced – different genes in different cells. As a result, a ball of originally identical cells begins to differentiate into distinct cell types, partly under the influence of those elusive chemical morphogens that can evidently control gene switching. By the time the embryo is fully formed, the differentiation process has created all the different cell types needed.*

All cells in your body have the *same* DNA, yet a skin cell is different from a liver cell is different from a brain cell. The information in DNA is referred to as the *genotype* and the actual physical cell is called the *phenotype*. So one genotype can generate many different phenotypes. Fine. But how do liver cells gather in the liver, brain cells in the brain, and so on – the cellular equivalent of 'birds of a feather flock together'? Most of what is known comes from the study of the fruit fly

---

* Some cells, pluripotent stem cells, remain only partially differentiated and have the potential to become cell types of different varieties.

Drosophila. Some of the morphogens are responsible for causing undifferentiated cells to differentiate into the various tissue types – eyes, gut, nervous system, and so on – in designated locations. This establishes a feedback loop between cell differentiation and the release of other morphogens in different locations. Substances called growth factors (I mentioned one called EGF earlier in this chapter) accelerate the reproduction of cells in that region, which will alter the local geometry via differential growth. This hand-wavy account is easy to state, but not so easy to turn into a detailed scientific explanation, in large part because it depends on the coupling between chemical networks and information-management networks, so there are two causal webs tangled together and changing over time. Added to all this is growing evidence that not just chemical gradients but physical forces – electric and mechanical – also contribute to morphogenesis. I shall have more to say on this remarkable topic in the next chapter.

Curiously, Alan Turing took an interest in the problem of morphogenesis and studied some equations describing how chemicals might diffuse through tissue to form a concentration gradient of various substances, reacting in ways that can produce three-dimensional patterns. Although Turing was on the right track, it has been slow going. Even for those morphogens that have been identified, puzzles remain. One way to confirm that a candidate chemical does indeed serve as a specific morphogen is to clone the cells that make it and implant them in another location (these are referred to as ectopic cells) to see if they produce a duplicate feature in the wrong place. Often, they do. Flies have been created with extra wings, and vertebrates with extra digits. But even listing all the substances that directly affect the cells immersed in them is only a small part of the story. Many of the chemicals diffusing through embryonic tissue will not affect cells directly but will instead act as signalling agents to regulate other chemicals. Untangling the details is a huge challenge.

A further complicating factor is that individual genes rarely act alone. As I have explained, they form networks in which proteins expressed by one gene can inhibit or enhance the expression of other genes. The late Eric Davidson and his co-workers at the California Institute of Technology managed to work out the entire wiring diagram (chemically speaking) for the fifty-odd gene network that regulates the early-stage development of the sea urchin (it was this lowly animal that attracted the attention of Dreisch a century ago). The Caltech group then programmed a computer, put in the conditions corresponding to the start of development and ran a simulation of the network dynamics step by step, with half-hour time intervals between them. At each stage they could compare the computer model of the state of the circuit with the observed stage of development of the sea urchin. Hey presto! The simulation matched the actual developmental steps (confirmed by measuring the gene expression profile). But the Davidson team went beyond this. They considered the effects of tweaking the circuitry to see what would happen to the embryo. For example, they performed experiments knocking out one of the genes in the network called *delta*, which caused the loss of all the non-skeletal mesoderm tissue – a gross abnormality. When they altered the computer model of the network in the corresponding way, the results precisely matched the experimental observations. In an even more drastic experiment, they injected into the egg a strand of mRNA that repressed the production of a critical enzyme called Pmar1. The effect was dramatic: the whole embryo was converted into a ball of skeletogenic cells. Once again, the computer model based on the circuit diagram described the same major transformation.

The various examples I have given illustrate the power and scope of 'electronic thinking' in tracking the flow of information through organisms and in linking it to important structural features. One of the most powerful aspects of the concept of information in biology is that the same general ideas often apply on all scales of size. In his visionary essay Nurse writes, 'The principles and rules that underpin how information is managed may share similarities at these different levels even though their elements are completely different . . . Studies at higher system levels are thus likely to inform those at the simpler level of the cell and vice versa.'[23]

So far, I have considered the patterns and flow of information at the molecular level in DNA, at the cellular level in the cell cycle of yeast, in the development of form in multicellular organisms, and in communities of organisms and their social organization. But when

Schrödinger conjectured his 'aperiodic crystal' he was focusing on *heritable* information and how it could be reliably passed on from one generation to the next. To be sure, information propagates in complex patterns within organisms and ecosystems, but it also flows vertically, cascading down the generations, providing the foundation for natural selection and evolutionary change. And it is here, at the intersection of Darwinism and information theory, that the magic puzzle box of life is now springing its biggest surprises.

4

Darwinism 2.0

'Nothing in biology makes sense except in the light of evolution.'

– Theodosius Dobzhansky[1]

'Nature, red in tooth and claw.' Alfred Tennyson penned these evocative words at the dawn of the Darwinian age. Understandably, scientists and poets of the day were wont to dwell on the brutality of natural selection as manifested in the arms races of bodily adaptations, be they the razor-sharp teeth of the shark or the tough defensive shell of the tortoise. It is easy to understand how evolution may select for bigger wings, longer legs, keener vision, and so on in the relentless struggle for survival. But bodies – the hardware of life – are only half the story. Just as important – indeed, more important – are the shifting patterns of information, the command-and-control systems, which constitute the software of life. Evolution operates on biological software just as it does on hardware; we don't readily notice it because information is invisible. Nor do we notice the minuscule demons that shunt and process all this information, but their near-thermodynamic perfection is a result of billions of years of evolutionary refinement.[2]

There is an analogy here with the computer industry. Thirty years ago personal computers were clunky and cumbersome. Innovations like the mouse, the colour screen and compact batteries have made computers far more efficient and convenient, as a result of which sales have soared. The capitalist version of natural selection has thus led to a vast growth in the population numbers of computers. But alongside the hardware innovations there has been an even more impressive advance in computer software. Early versions of Photoshop or Power-Point, for example, are a pale shadow of those currently available. Above all, the speed of computers has increased vastly, while the cost has plummeted. And software improvements have contributed at least as much as hardware embellishments to the success of the product.

It took a century following the publication of Darwin's theory for the informational story of life to enter the evolutionary narrative. The field of bioinformatics is now a vast and sprawling industry, accumulating staggering amounts of data and riding high on hyperbole. The publication in 2003 of the first complete human genome sequence, following a mammoth international effort, was hailed as a game-changer for biology in general and medical science in particular. Although the importance of this landmark achievement should not be diminished, it soon became clear that having complete details of a genome falls far short of 'explaining life'.

When Darwin's theory of evolution was combined with genetics and molecular biology in the mid-twentieth century, in what is termed the 'modern synthesis', the story seemed deceptively simple. DNA is a physical object; copying it is bound to incur random errors, providing a mechanism for genetic variation on which natural selection can act. Make a list of the genes and the function of the proteins they code for and the rest will be mere details.

About twenty years ago this simplistic view of evolution began to crumble. The road from an inventory of proteins to functional three-dimensional anatomy is a long one and the protein 'parts list' provided by the genome project is useless without the 'assembly instructions'. Even today, in the absence of foreknowledge, nobody can predict from a genomic sequence what the actual organism might look like, let alone how random changes in the genome sequence would translate into changes in phenotype.

Genes make a difference only when they are expressed (that is, switched on), and it is here, in the field of gene control and management, that the real bioinformatics story begins. This emerging subject is known as epigenetics, and it is far richer and more subtle than genetics in isolation. More and more epigenetic factors which drive the organization of biological information patterns and flows are being discovered. The refinement and extension of Darwinism that is now emerging – what I am calling Darwinism 2.0 – is yielding an entirely new perspective on the power of information in biology, ushering in a major revision of the theory of evolution.

ELECTRIC MONSTERS

'There is more to heredity than genes.'

– Eva Jablonka[3]

'It came from space! Two-headed flatworm stuns scientists.'[4] So proclaimed a British online publication in June 2017. The subject of the article, which inevitably involved 'baffled boffins', was the appearance of monsters in the International Space Station. The monsters didn't invade the station; they emerged as part of an experiment to see how lowly flatworms got on in orbit if their heads and tails were chopped off in advance. It turns out they got on very well. One in fifteen came home with two heads in place of the one they had lost.[5]

The space worms are just one rather dramatic example of the exploding field of epigenetics. Loosely defined, epigenetics is the study of all those factors which determine the forms of organisms that lie beyond their genes (see Box 9). The two-headed worms are genetically identical to their more familiar cousins, but they look like a different species. Indeed, two-headed worms reproduce and beget more two-headed worms. No wonder the boffins were baffled. In this case the chief boffin was Michael Levin of Tufts University, who happens to be a collaborator with our research group at ASU.

To put the worm work in context, recall from the previous chapter how the development of the embryo (morphogenesis) provides a graphic example of the power of information to control and shape the form of the organism, although many of the actual mechanisms at work remain puzzling. I explained that the information needed to build and operate an organism lies to a great extent in the system's ability to switch genes on or off and to modify proteins after the genetic instructions are translated. The regulation of information flow via chemical pathways – involving molecules such as the methyl group, histone tails and micro-RNA (see Box 9) – and the coupling of this gene-switching repertoire to a plethora of shifting chemical patterns is as yet dimly understood. Epigenetics thus opens up a vast universe of combinations and possibilities. I mentioned how the diffusion of specialized molecules called morphogens play an important role in controlling the unfolding dynamics of development, but that turns out to be only part of the story. In the last few years it has become clear that another physical mechanism could be even more significant in morphogenesis. Known as *electro-transduction*, it deals with changes to an organism's form arising from electrical effects.

Box 9: Beyond the gene

Genes are switched on and off as needed through the life of an organism. There are many ways a gene can be silenced. A common method is methylation, in which a small molecule of the methyl group becomes attached to the letters C in a gene and physically blocks the gene readout mechanism. Another is RNAi, a tiny RNA fragment just twenty-something letters long discovered serendipitously by botanists who were trying to make prettier flowers. In this mechanism, the gene is read out from DNA as usual, but RNAi (i stands for interference) mugs the messenger RNA while it is busy taking its data read-out to the ribosome and chops it in two, thus (somewhat brutally) junking the message. In complex organisms, genes can be smothered by being buried in highly compacted regions of chromatin.

In addition to gene switching, several other variables are at play. An expressed gene may produce a protein that subsequently becomes modified in some way. For example, proteins called histones assemble into little yo-yo structures known as nucleosomes around which DNA wraps itself. A single human chromosome may contain hundreds of thousands of nucleosomes. The yo-yos are not just structural elements but implicated in gene regulation. A variety of small molecules can become attached to the histones to make tails; there is some evidence that these molecular tags themselves form a code. Also, the spacing between nucleosomes along the DNA is neither regular nor random, and it seems that the positioning patterns contain important information in their own right. All these variables are very complex and the details are still not completely understood, but it is clear that modifications made to proteins *after* their manufacture are important regulatory elements in the cell's information-management system. A further complication is that a 'gene' is not necessarily a continuous segment of DNA but may be made of several pieces. As a result, the mRNA read-out has to be cut and spliced to assemble the components correctly. In some cases, there is more than one splicing, meaning that a single stretch of DNA can code for several proteins at once; which protein is expressed depends on the specifics of the splicing operation, which is itself managed by other genes and proteins . . . and so on.

Perhaps the biggest source of variety comes from the fact that, at least in complex organisms like animals and plants, the vast majority of DNA does not consist of genes coding for proteins anyway. The purpose of this 'dark sector' of DNA remains unclear. For a long while much of the non-coding DNA segments were dismissed as junk, as serving no useful biological function. But increasingly there is evidence that much of the 'junk' plays a crucial role in the manufacture of other types of molecules, such as short strands of RNA, which regulate a whole range of cellular functions. Cells are beginning to look like bottomless pits of complexity. The discovery of all these causal factors which are not located on the actual genes is part of the field known as epigenetics. It seems that epigenetics is at least as important as genetics as far as biological form and function are concerned.

In a pale echo of Frankenstein, it turns out that electricity is indeed a life force, but not quite in the way Mary Shelley (or rather Hollywood's version of her story) imagined. Most cells are slightly electrically charged. They maintain this state by pumping positively charged ions (mostly protons and sodium) from inside to outside through the membrane that encloses the cell, creating a net negative charge. Typical potential differences across the membranes are between 40 and 80 millivolts. Although that doesn't seem high, the membrane is so thin that this small voltage gradient represents an enormous local electric field – greater than those found near the Earth's surface during a thunderstorm – and it's actually possible to measure it. By using voltage-sensitive fluorescent dyes researchers can make pictures of the field patterns.

In a series of spectacular experiments at Tufts University, Michael Levin – he of the space worms – has demonstrated that electric patterning is important in sculpting the final morphology of an organism as it develops. Variations in voltage across large areas of the body serve as 'pre-patterns' – invisible geometrical scaffolds that drive downstream gene expression and thereby affect the path of development. By manipulating the electric potential differences across selected cells, Levin can disrupt the developmental process and create monsters to order – frogs with extra legs and eyes, worms with heads where tails should go, and so on.*

One series of experiments focuses on tadpoles of *Xenopus*, the African clawed frog. Normal frog embryos develop a characteristic pattern of pigmentation after a fraction of cells in the mid-region of the head and trunk begin to produce melanin. Levin treated the tadpoles with ivermectin, a common anti-parasite agent that electrically depolarizes cells by changing the flow of ions between the cell and its surroundings. Altering the electrical properties of so-called instructor cells had a dramatic effect, causing the pigmented cells to go crazy, spreading cancer-like into distant regions of the embryo. One perfectly normal tadpole developed a metastatic melanoma entirely from the electrical disruption, in the absence of any carcinogens or mutations. That tumours may be triggered purely epigenetically contradicts the prevailing view that cancer is a result of genetic damage, a story that I shall take up later in the chapter.

---

* Although electricity is key, the morphogenetic field here is not an electric field in the normal sense, extending across the developing tissue. Instead, it is a field of electric cell polarization. 'Polarization' is the term given to describe the voltage difference across the cell's membrane. If that voltage drop varies from cell to cell and place to place, it may be said that there is an electric polarization field spread throughout the developing tissue. Physicists will recognize that the polarization is a scalar field, whereas electric fields are vector fields.

All this was remarkable enough. But an even bigger surprise lay in store. In a different experiment at Tufts University, devised by Dany Adams, a microscope was fitted with a time-lapse camera to produce a movie of the shifting electric patterns during the development of *Xenopus* embryos. What it showed was spectacular. The movie began with a wave of enhanced electrical polarization sweeping across the entire embryo in about fifteen minutes. Then various patches and spots of hyperpolarization and depolarization appeared and became enfolded as the embryo reorganized its structure. The hyperpolarized regions marked out the future mouth, nose, ears, eyes and pharynx. By altering the patterns of these electrical domains and tracing how the ensuing gene expression and face patterning changed, the researchers concluded that the electrical patterns *pre-figure* structures scheduled to emerge much later in development, most strikingly in the face of the frog-to-be. Electrical pre-patterning appears to guide morphogenesis by somehow storing information about the three-dimensional final form and enabling distant regions of the embryo to communicate and make decisions about large-scale growth and morphology.

Embryo development is a dramatic example of biological morphogenesis. Another is regeneration. Some animals can regrow their tails, even entire limbs, if they are lost for some reason. And sure enough, there's an electrical story here too. Levin's experimental creature of choice is a type of flatworm called a planarian (the 'space worm' species). These tiny animals have a head at one end with eyes and a brain to go with it, and a tail at the other end. Planaria are a favourite with teachers because if they are chopped in two they don't die; instead, writes Levin,

> the wound on the posterior half builds a new head, while the wound on the anterior half makes a tail. Two completely different structures are formed by cells that, until the cut occurred, were sharing all aspects of the local environment. Thus, still poorly understood long-range signals allow the wound cells to know where they are located, which direction the wound is facing and what other structures are still present in the fragment and do not need to be replaced.[6]

Levin discovered there is a distinctive electric pattern throughout the cut fragment, as there is indeed around wounds generally. Levin used drugs called heptanol and octanol, which sound like rocket fuel but serve to interfere with the ability of cells to communicate electrically with each other and hence modify the activity of the bioelectric circuit controlling how the tissues around the wounds decide their identity. By this means he was able to get a severed head to grow not a tail but another head, thus creating a two-headed, zero-tailed worm (see Fig. 12). Likewise, he can make two-tailed worms with no head. (He can even make worms with four heads or four tails.) The biggest surprise comes if the experimenters chop off the aberrant supernumerary head of a two-headed worm. You might expect this to rid the worm of any further two-headed aspirations, but it turns out that if the remainder of the worm is then cut in half two new two-headed worms are made! This is a dramatic example of *epigenetic inheritance* at work (see Box 10). The key point is that all these monster worms have

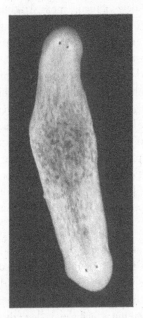

Fig. 12. Two-headed worm created by Michael Levin at Tufts University using manipulation of electrical polarity. The worm reproduces other two-headed worms when bisected, as if it is a different species, although it has the same DNA as normal one-headed worms.

identical DNA sequences yet dramatically different phenotypes. A visitor from Mars would surely classify them as different species based on their morphology. Yet, somehow, the physical properties of the organism (in this case, the stable states of the electric circuits) convey altered morphological information from one generation to the next.

Which brings up two important questions: *where* is the morphological information stored in these creatures, and how is it passed on between generations? Obviously, the information is not in the genes, which are identical. The DNA alone does not directly encode shape (anatomical layout) or the rules for repairing that shape if damage occurs. How do tissues know to keep rebuilding, say, the head of a planarian and stop when the right size has been completed? The standard reductionist explanation is to attribute the organism's regenerative capabilities to a set of inherited genetic instructions, along the lines of 'what to do if you are chopped in two: grow a new tail if you have lost your original', and so on. But given that a two-headed planarian has the *same* set of genes as a normal planarian, how does a newly bisected two-headed worm tell its exposed stump 'make a head' in defiance of the normal protocol of 'make a tail'? What epigenetic apparatus, exactly, is adapted by the momentary electrical tinkering yet remains locked in place for generation after generation of monsters even after Levin's rocket fuel is removed?

The biggest problem here is not untangling the story of what proteins are where but how the system *as a whole* processes information on size, shape and topology over a scale much larger than any single cell. What is needed is a top-down view that focuses on information flow and mechanisms for encoding the shapes of large and complex structures. So far, however, that code, or the nature of the signals conveying the building and repair instructions, remains shadowy. One way forward is to imagine that there is some sort of 'information field' permeating the organism, which, after Levin and his collaborators have adulterated it, somehow embeds details about the large-scale properties of the monster-in-waiting, including its three-dimensional form. Just how that might work is anybody's guess. The way Levin expresses it, there is a pre-existing 'target morphology' that guides a variety of shape-modulating signals and is stored, interpreted and implemented by a combination of chemical, electrical and mechanical processes acting in concert:

> A 'target morphology' is the stable pattern to which a system will develop or regenerate after perturbation. Although not yet understood mechanistically, regeneration ceases when precisely the right size structure has been rebuilt, indicating a coordination of local growth with the size and scale of the host.[7]

Understanding the growth of complex forms in biology has enormous medical implications, ranging from birth defects to cancer. If these forms are mediated at least in part by electric patterning, or indeed by any encoding that we can learn to rewrite (and let cells build to specification), there is scope for correcting and controlling pathologies. The holy grail of regenerative medicine is to be able to regrow entire organs. The human liver will in fact regrow to its normal size following surgical resectioning. Again, how it knows the final shape and size is puzzling. If similar regeneration could be extended to nerves, cranio-facial tissue and even limbs, the applications would be stupendous. But achieving these goals requires a much better understanding of living systems as cohesive, computational entities that store and process information about their shape and their environment. Above all, we need to discover how informational patterns – electrical, chemical and genetic – interact and translate into specific phenotypes.

Electro-transduction is just one example of how physical forces can affect gene expression. Mechanical pressure or sheer stress acting on the cell as a whole can sometimes produce changes in the cell's physical properties or behaviour. A well-known example is contact inhibition. Cells in a Petri dish will happily divide if they are treated well and receive nutrients, but division will stop if the cells become claustrophobic, such as when the colony pushes up against a boundary and the population becomes overcrowded. Cancer cells turn off contact inhibition. They also undergo drastic changes in their shape and stiffness when they leave the primary tumour and spread around the body. Another example: when a stem cell is placed next to a hard surface, it will express different genes than when embedded in soft tissue, affecting the type of cell it differentiates into, a phenomenon of obvious importance to embryogenesis. There is a popular aphorism in the cancer research community: 'What a cell touches determines what a cell does.' The mechanism behind these sorts of phenomena is known as mechano-transduction, meaning that an external mechanical signal – a gross physical force – is transduced into altered gene expression in response.[8]

The two-headed space worms provide a striking illustration of mechano-transduction in zero-gravity conditions. Another space wonder comes from my own university and the experiments of Cheryl Nickerson. She has been working with NASA to study changes in the gene expression of microbes when they go into Earth orbit. Even the humble salmonella bacterium can somehow sense it is floating in space and changes its gene expression accordingly.[9] The finding has obvious implications for astronauts' well-being, because a nasty bug that might be held in check on Earth may make someone sick in space. Related to this is the fact that humans normally carry about a trillion microbes around inside them, many in the gut, forming what is called their microbiome; it plays an important role in human health. If there are changes in gene expression within the microbiome due to long periods in zero- or low-gravity conditions, it could become a serious obstacle to long-term spaceflight.[10]

Let me mention a couple more intriguing discoveries to finish up this section. Salamanders are well known for their ability to regenerate entire limbs. It turns out that if the sacrificed leg has cancer, and is severed in mid-tumour, the new leg is cancer-free. Evidently, the limb morphology, somehow encoded in the stump, is re-programmed to make a healthy leg. This runs counter to the conventional wisdom that rapid cell proliferation – a feature of limb regeneration – is a cancer risk (cancer is sometimes described as 'the wound that never heals'). Indeed, many studies have also shown the ability of embryos to tame aggressive cancer cells. Another oddity concerns deer antlers, which drop off and regrow every year. In some species of deer, if you cut a notch in the antlers, next year's regrown antlers come complete with an ectopic branch (tine) at that same location.* Where, one wonders, is the 'notch information' stored in the deer? Obviously not

---

* There have been no tests yet to see whether baby deer inherit the change.

in the antlers, which drop off. In the head? How does a deer's head know its antler has a notch half a metre away from it, and how do cells at the scalp store a map of the branching structure so as to note exactly where the notch was? Weird! In the magic puzzle box of life, epigenetic inheritance is one of the more puzzling bits of magic.

Box 10: Epigenetic inheritance

Like the proverbial farmer's wife, the German evolutionary biologist August Weismann cut off the tails of mice over many generations, but he never succeeded in producing a tailless mouse – a blow to Lamarck's theory of evolution by the inheritance of acquired characteristics. However, the recent surge in the study of epigenetics is painting a more nuanced picture. Within the body, a cell's type is conserved when the cell divides: if, say, a liver cell replicates, it makes two liver cells, not a liver cell and a skin cell. So the epigenetic markers (for example, methylation patterns) that determine gene expression ('Thou shalt be a liver cell') will be passed on to the daughter cells. But what about epigenetic changes passed from one generation *of the whole organism* to the next, for example, from mother to son? That is a very different matter; if it occurred, it would strike at the very basis of Darwinian evolution. There is not supposed to be any mechanism for changes to an organism's body to get into its germ line (sperm and eggs) and affect its offspring. Nevertheless, evidence for intergenerational epigenetic inheritance has long been staring biologists in the face. When a male donkey is crossed with a female horse it produces a (sterile) mule. If a female donkey is crossed with a male horse, the result is a hinny. Mules and hinnies are genetically identical but very different-looking animals: they are epigenetically distinct, and so they must carry epigenetic determinants that depend on the sex of their parents. Other examples have been discovered where genes from the parents are imprinted with epigenetic molecular marks that manage to get into the germ cells and survive the reproductive process. Moreover, botanists know many cases where epigenetic changes accumulated during a plant's life are passed on to the next and subsequent generations. Even in humans some studies have uncovered hints of something similar. One of these involved Dutch families who suffered near-starvation in the Second World War because they were bypassed by the Allied advance and unable to receive food. Children of the survivors were born with lower than average body weight and stayed below average height all their lives. More surprisingly, their own children seem to be smaller too.

So where, exactly, *is* the epigenome? Genes are physical objects with a definite location in the cell: a specific gene lies at a certain position on DNA. You can see them with a microscope. When it comes to epigenetics, however, there is no 'epigenome' in the same physical sense, no well-defined object at a specific location in the cell. Epigenetic information processing and control is distributed throughout the cell (and perhaps beyond the cell too). It is global, not local, as is the case with the cellular analogue of von Neumann's supervisory unit I mentioned in the previous chapter.

FLIRTING WITH LAMARCKISM

'Chance favors the prepared genome.'

– *Lynn Caporale*[11]

Decades before Darwin released *On the Origin of Species*, a French biologist published a very different theory of evolution. His name was Jean-Baptiste Lamarck. The centrepiece of Lamarckian evolution is that characteristics acquired by an organism during its lifetime can be inherited by its offspring. Thus, if an animal strives in this or that manner (tries to run faster, reach higher . . .) in its relentless struggle for survival, its progeny will come with an inherited slight improvement (a bit swifter, a bit taller). If this theory is correct, it would provide a mechanism for fast-paced purposeful change directed towards betterment. My mother often remarked that she could really use another pair of hands when doing housework. Imagine if, as a result of this need, her children had been born with four arms! By contrast, Darwin's theory asserts that mutational changes are blind; they have no link to the circumstances or requirements of the organism that carries them. If a rare mutation confers an advantage, then that's pure dumb luck. There is no directional progress, no systematic inbuilt mechanism for improvement.

Evolution would certainly work much faster and more efficiently if nature engineered just the right mutations to help out, in the way Lamarck envisaged. Nevertheless, biologists long ago dismissed the idea as too much like the guiding hand of God, preferring instead to appeal to random chance as the sole explanation for variation. And that's the way it was for many decades. Now, however, doubts have crept in. Levin's monster worms are surely a clear example of the inheritance of acquired characteristics – acquired in this case by laboratory butchery. Many other examples are known. So has the time come to abandon Darwinism and embrace Lamarckism?

Nobody can deny that natural selection encourages the survival of the fittest. Organisms show variations, and nature selects the fitter. But there have always been niggling worries. Nature can only work with the variants it has, and a fundamental question is how those variants arise. Survival of the fittest maybe, but what about the *arrival* of the fittest, as the Dutch botanist Hugo de Vries dubbed it a century ago? In biology, remarkable innovations with far-reaching consequences abound: photosynthesis, the bony skeleton of vertebrates, avian flight, insect pollination, neural signalling, to name but a few. The problem of how life generates so many ingenious solutions to survival problems is today the subject of lively investigation.[12] If something works well, random changes are likely to make it worse, not better. Even with a duration of 3 or 4 billion years, is it possible that so much organized complexity – eyes, brains, photosynthesis – has arisen *just* from random variation and natural selection?[13]

Over the years many scientists have expressed scepticism. 'A simple probabilistic model would not be sufficient to generate the fantastic diversity we see,'[14] wrote Wolfgang Pauli, a quantum physicist and contemporary of Schrödinger. Even distinguished biologists have expressed their doubts. Theodor Dobzhansky wrote: 'The most serious objection to the modern theory of evolution is that since mutations occur by "chance" and are undirected, it is difficult to see how mutation and selection can add up to the formation of such beautifully balanced organs as, for example, the human eye.'[15] Many of these problems would evaporate if some vestige of Lamarckian evolution were at work.

In 1988 a group of Harvard biologists claimed they had witnessed a clear-cut example of the propitious arrival of the fittest. A team led by John Cairns made the provocative claim 'Cells may have mechanisms for choosing which mutations will occur.'[16] *Choosing?* Published in *Nature*, and coming from a highly respectable laboratory, the announcement was received with consternation. To understand their experiments, recall from p. 88 that the bacteria *E. coli* like to eat glucose but can flip a switch to enable them to metabolize less-tasty lactose if pressed. Cairns' group worked with a mutant strain of *E. coli* that was unable to process lactose, and then challenged them with a lactose-only diet. They observed that some of these starving bacteria spontaneously mutated to the lactose-utilizing form. In itself, this is no threat to orthodox Darwinism, so long as the said mutations arise as lucky flukes. But when the Harvard team worked out how likely that was, they concluded that their bacteria had uncanny rates of success, beating the raw odds by a huge margin. The researchers wondered, 'Can the genome of an individual cell profit by experience?'[17] just as Lamarck had proposed. They hinted that the answer might be yes, and that they were dealing with a case of mutations '"directed" toward a useful goal'.

Reacting to the furore, Cairns did some follow-up experiments and backtracked on the more sensational aspects of the claim. But the genie was out of the bottle, and there ensued a surge of experiments by his and other groups. A lot of *E. coli* suffered glucose deprivation. When the dust settled, this is what emerged. Mutations are *not* random: that part is correct. Bacteria have mutational hotspots – specific genes that mutate up to hundreds of thousands of times faster than average. This is handy if it is advantageous for the bacteria to generate diversity. A case in point is when they invade a mammal and have to do battle with the host's immune system. Bacteria have identifying surface features that act a bit like a soldier's uniform. The host's immune system recognizes the pathogen from the details of its coat. A bacterium that can keep changing uniform will obviously have a survival advantage, so it makes good Darwinian sense for the 'uniform genes' to be highly mutable. For circumstances such as this, bacteria have evolved certain 'contingency genes' that are more mutable than others, implying a greater likelihood of mutations arising in these genes. Within that contingency, however, it's still a hit-and-miss affair. There is no evidence for bacteria 'choosing' specific mutations to order, as Cairns originally hinted.

A more arresting example concerns bacteria that can selectively switch on elevated mutation rates in just the right genes to get them out of trouble. Barbara Wright of the University of Montana looked at mutants of poor old E. coli that possessed a defective gene, one that codes for making a particular amino acid.[18] You and I usually get our amino acids from food, but if we're starving our cells can make their own. Same with bacteria. What Wright wondered was how a starving bacterium with a faulty amino acid gene would respond. The bacterium gets the signal 'need amino acids now!' but the defective gene cranks out a flawed version. Somehow, the cell senses this danger and turns up the mutation rate of that specific gene. Most mutations make things worse. But among a colony of starving bacteria there is now a good chance that one of them will get lucky and suffer just the right mutation to repair the defect, and that cell saves the day. It gets the bacterial equivalent of another pair of hands. The term used for this biased mutation is 'adaptive', because it makes the organism better adapted to its environment.

A long-time pioneer of adaptive mutations is Susan Rosenberg, now at Baylor College of Medicine in Houston, Texas. Rosenberg and her colleagues also set out to get to the bottom of how starving bacteria manage to mutate their way to a culinary lifeline with such uncanny panache. They focused on the repair of double-strand breaks in DNA – a never-ending chore required so that cells can carry on their business as usual.[19] There are various methods used to patch such a gap, some high quality, others less so. Starving bacteria, Rosenberg discovered, can switch from a high-fidelity repair process to a sloppy one. Doing so creates a trail of damage either side of the break, out as far as 60,000 bases or more: an island of self-inflicted vandalism. Rosenberg then identified the genes for organizing and controlling this process. It turns out they are very ancient; evidently, deliberately botching DNA repair is a basic survival mechanism stretching back into the mists of biological history. By generating cohorts of mutants in this manner, the colony of bacteria improves its chances that at least one daughter cell will accidentally hit on the right solution. Natural selection does the rest. In effect, the stressed bacteria engineer their own high-speed evolution by generating genomic diversity on the fly.

Is there any hint from Rosenberg's experiments that these wily bacteria can also generate the 'right' mutations with better than chance, as Cairns originally implied? Do the fittest 'arrive' with prescient efficiency? This is not a simple yes-or-no issue. It's true that the cells don't adopt a scattergun approach in their mutational rampage: the elevated mutations are not distributed uniformly across the genome. Rosenberg confirmed, however, the existence of certain favoured hotspots which are more likely than chance to house the genes needed to evolve out of trouble. But unlike the highly focused mechanism that Barbara Wright discovered, which targets specific faulty genes expressing themselves badly, Rosenberg's mutations affect *all* genes in the hotspot regions indiscriminately, regardless of whether they are struggling to churn out proteins or just sitting idle. In that sense it is a more basic yet also more versatile mechanism.

Here is an analogy. Imagine being trapped inside a burning building. You guess there might be a window somewhere that will open and let you escape, but which one? Maybe there are dozens of windows. A really smart person would have figured out a fire escape procedure in advance, just in case. But you didn't. So what is the next smartest thing to do? It is of course to try each of the windows one by one. In the absence of any other information, a random sampling procedure is as good as any. A really dumb thing to do would be to scurry around *totally* at random, going into cupboards or ducking under beds. *Targeted* randomness is more efficient than complete randomness. Well, bacteria are not super-smart but nor are they really dumb: they concentrate their chances where it is most likely to do some good.

How did all this mutational magic come to exist? In retrospect, it's not too surprising. Clearly, evolution will work much better if the mechanisms involved are flexible and can *themselves* evolve – what is often called the evolution of evolvability. Long, long ago, cells that retained an ability to evolve their way out of trouble would have been at an advantage. An evolution-boosting mechanism that is switched on when conditions demand and dialled back when times are great is a boon. The adaptive response to stress* is almost certainly an ancient mechanism (really a set of mechanisms involving a spectrum of processes, from haphazard to focused and directed) that evolved for good biological purposes. The biologist Eva Jablonka describes adaptive mutations as an 'informed search'. She concludes, 'The cell's chances of finding a mutational solution are enhanced because its evolutionary past has constructed a system that supplies intelligent hints about where and when to generate mutations.'[20] It is important to understand that this is not a falsification of Darwinism but an elaboration of it. This is Darwinism 2.0. Biochemist Lynn Caporale writes: 'Rejecting entirely random genetic variation as the substrate of genome evolution is not a refutation, but rather provides a deeper understanding, of the theory of natural selection of Darwin and Wallace.'[21] The refinement that emerges from these recent experiments displaying a Lamarckian flavour is that nature selects not just the fittest organisms but the *fittest survival strategies* too.

The foregoing ideas illustrate how organisms use information from the past to chart the future. This information is inherited both over deep time (for example, the contingency genes I discussed on p. 124) and epigenetically, from the previous generation. Life may therefore be described as an informational learning curve that swoops upwards. Organisms do not have to proceed by trial and error and 'reinvent the wheel' at each generation. They can profit from life's past experience. This progressive trend stands in stark contrast to the second law of thermodynamics, which tells a story of degeneration and decay.

DEMONS IN THE GENES

Surprising though adaptive mutations may be, they still imply that genomes are the passive victims of randomly inflicted externally

---

* The word 'stress' here, and in what follows in this chapter, doesn't of course refer to a mental state but to circumstances in which a cell or organism is threatened or challenged in some way, for example, by starvation or wounding.

generated blows or blunders, albeit with rigged odds. It's still a chancey business. But suppose cells in trouble didn't have to rely on external forces at all to cause mutations? What if they could *actively manipulate their own genomes*?

Actually, it is clear that they do. Sexual reproduction involves several slicing-and-dicing genomic reconfigurations, some random, some supervised. Many intermingling methods are out there, each of which involves cells shuffling their own DNA in a carefully arranged manner. And sex isn't the only example. Correcting the errors that occur during DNA replication requires another set of genomic management operations. Most of the primary damage to DNA, for example by radiation or thermal disruption, never makes it to the daughter cells because it is repaired first. Human DNA would suffer devastating mutational damage, estimated at an overall 1 per cent copying error rate per generation, without all the in-house, high-tech proofreading, editing and error correcting which reduces the net mutation rate to an incredible one in 10 billion. So cells are able to monitor and actively edit their own genomes to a high degree of fidelity in an attempt to maintain the status quo.

But now we encounter a fascinating question: can cells actively edit their genomes to *change* the status quo? Decades before the work of Cairns and Rosenberg this question was investigated in a series of remarkable experiments by the distinguished botanist and cell biologist Barbara McClintock. Starting in her student days in the 1920s, she experimented with maize plants and established many of the basic properties of chromosome structure and organization we know today, for which she subsequently received a Nobel Prize in Physiology or Medicine – the first woman to win the prize unshared in that category. With the help of a basic microscope, McClintock looked to see what happened to the chromosomes of the maize plants when they were exposed to X-rays. What she reported caused such a ballyhoo and attracted so much scepticism that in 1953 she felt moved to stop publishing her data. What was uncontentious were her observations that chromosomes break into fragments when irradiated. But the big surprise was the fact that the pieces could rejoin again, often in novel arrangements. Humpty-dumpty could be reassembled in a baroque sort of way. Chromosome reorganization writ large might seem lethal, and it often was. But it was not always so: in some cases, the mutant plants went on to replicate their grossly modified genomes. Crucially, McClintock found that the large-scale mutations were far from random; it appeared that the maize cells had a contingency plan for the day their genomes were smashed. Even more amazingly, if the plants were stressed, for example by infection or mechanical damage, *spontaneous* chromosome breakage could occur without the benefit of X-ray disruption; the broken ends were rejoined after the chromosome replicated. In 1948 McClintock made her most startling discovery of all. Segments of chromosomes could be transposed – switch places on the genome – a phenomenon popularly known as 'jumping genes'. In the maize plants this produced a mosaic coloured pattern.

Today, genomic transpositions are recognized as widespread in evolution. It has been estimated that up to half the human genome has undergone such genetic gymnastics. Cancer researchers are also very familiar with transpositions. A much-studied example is the Philadelphia chromosome (after where it was discovered) which can trigger leukaemia in humans; it involves a chunk of chromosome 9 being transposed for a chunk of chromosome 22. In certain late-stage cancers, chromosomes can become so deranged as to be almost unrecognizable, with wholesale rearrangements, including entire chromosome duplications and stand-alone fragments replacing the orderly arrangement found in healthy cells. An extreme case is called chromothripsis, in which chromosomes disintegrate into thousands of pieces and rearrange themselves into scrambled monsters.

In spite of the grudging acknowledgement that McClintock was right, her results remain disturbing because they imply that the cell can be an active agent in its own genomic change. She herself evidently thought as much. On the occasion of her Nobel Prize, awarded for 'the discovery of mobile genetic elements', she had this to say:

> The conclusion seems inescapable that cells are able to sense the presence in their nuclei of ruptured ends of chromosomes and then to activate a mechanism that will bring together and then unite these ends, one with another ... The ability of a cell to sense these broken ends, to direct them toward each other, and then to unite them so that

the union of the two DNA strands is correctly oriented, is a particularly revealing example of the sensitivity of cells to all that is going on within them . . . A goal for the future would be to determine the extent of knowledge the cell has of itself, and how it utilizes this knowledge in a 'thoughtful' manner when challenged . . . monitoring genomic activities and correcting common errors, sensing the unusual and unexpected events, and responding to them, often by restructuring the genome. We know about the components of genomes that could be made available for such restructuring. We know nothing, however, about how the cell senses danger and instigates responses to it that often are truly remarkable.[22]

It turned out that transposition and mobile genetic elements were only the tip of the iceberg. When facing challenges, cells have many ways to 'rewrite' their genomes, just as computer programs have bugs removed or are upgraded to perform new tasks. James Shapiro, a collaborator with McClintock as a young man, has made a comprehensive study of the mechanisms involved. One of these is called reverse transcription, whereby RNA, which normally transcribes sequences *from* DNA, is sometimes able to write its own sequence *back into* DNA. Because there are many mechanisms for RNA sequences to be modified *after* they have transcribed the information from DNA, reverse transcription opens the way for cells to alter their own DNA via RNA modification. A specific reverse transcription gene that has been studied in detail is BC1 RNA, which is important in the neural systems of rodents.[23]

It is now recognized that diverse processes of reverse transcription have played a major role in evolution and may, for example, account for a large fraction of the genetic differences between humans and chimpanzees.

Nor is the backflow of information limited to RNA $\rightarrow$ DNA. Because genome repair is controlled by complex interactions in the cell, the decision 'to repair or not to repair' or 'how to repair' can depend on a variety of proteins that have been modified after their formation. The upshot is that proteins, and modifications they have acquired during the life cycle of the cell, can influence genomic content: the epigenetic tail wagging the genetic dog. In all, Shapiro has identified about a dozen different mechanisms whereby a cell, operating at a systems level, can affect the information content of its own DNA, a process he calls *natural genetic engineering*. To summarize the central dogma of neo-Darwinian biology, information flows from inert DNA to mobile RNA to functional proteins in a one-way traffic. To use a computer analogy, the Darwinian genome is a read-only data file. But the work of McClintock, Shapiro and others explodes this myth and shows it is more accurate to think of the genome as a read–write storage system.

The refinements of Darwinism I have described in this chapter go some way to explaining the puzzle of the arrival of the fittest. At present, there are collections of case studies hinting at different mechanisms, many with Lamarckian overtones, but as yet no systematic information-management laws or principles have been elucidated that govern these phenomena. However, it's tempting to imagine that biologists are glimpsing an entire shadow information-processing system at work at the epigenetic level. 'Nature's many innovations – some uncannily perfect – call for natural principles that accelerate life's ability to innovate . . .'[24] writes Andreas Wagner, an evolutionary biologist from Switzerland. 'There is much more to evolution than meets the eye . . . Adaptations are not just driven by chance, but by a set of laws that allow nature to discover new molecules and mechanisms in a fraction of the time that random variation would take.'[25] Kevin Laland, an evolutionary biologist at the University of St Andrews, is co-founder of what has been dubbed the 'Extended Evolutionary Synthesis'. 'It is time to let go of the idea that the genes we inherit are a blueprint to build our bodies,' he writes. 'Genetic information is only one factor influencing how an individual turns out. Organisms play active, constructive roles in their own development and that of their descendants, so that they impose direction on evolution.'[26]

Orthodox biologists are not taking this assault lying down. The heresy of Lamarckism is always guaranteed to inflame passions and the Extended Evolutionary Synthesis remains a contentious challenge, as is the claim that epigenetic changes can be passed down the generations. Just how much the 'purist' version of Darwinism needs to be adapted is controversial.[27] It's fair to say that the battle is far from over.

CANCER: THE HARSH PRICE OF
MULTICELLULARITY

Genomes can undergo profound changes not just on evolutionary timescales over millions of years but during the lifetime of an organism. The most dramatic example of the latter is provided by cancer, the world's number-two killer. Dreadful though this disease may be, it provides a fascinating window on our evolutionary past.

There is no hard-and-fast definition of cancer; instead it is characterized by about a dozen 'hallmarks'.[28] Advanced cancers in humans may display all or only some of the hallmarks. They include a surging mutation rate, unrestrained cell proliferation, disabling of apoptosis (programmed cell death), evasion of the immune system, angiogenesis (organization of a new blood supply), changes in metabolism and – the most well-known and medically problematic – a penchant for spreading around the body and colonizing organs remote from the site of the primary tumour, a process called metastasis.

Cancer is the most studied subject in biology, with over a million published papers in the last fifty years. It may therefore come as a surprise to the reader to learn that there is no agreement on what cancer is, why it exists and how it fits into the great story of life on Earth. Very little attention has been given to understanding cancer *as a biological phenomenon*, as opposed to a disease to be annihilated by any means at hand. Most of the gargantuan research effort across the world has been devoted to destroying cancer. Nevertheless, standard cancer therapy – a mix of surgery, radiation and chemical toxins – has changed little in decades.* Survival rates for all but a handful of cancer types have improved only modestly or not at all: life extension through chemotherapy is mostly a rearguard action against the inevitable, measured in weeks or months rather than years. This dismal state of affairs can't be blamed on lack of funding. The US government alone has spent $100 billion on cancer research

---

* Recently a fourth line of attack – immunotherapy – has received a lot of attention. It involves supercharging the body's immune system to destroy cancer cells. Early results are promising, but it is too soon to know if this technique will transform the field.

since President Nixon supposedly declared war on it in 1971, while charities and drug companies have poured in billions more.

Perhaps the lack of progress is because scientists are looking at the problem in the wrong way? Two common misconceptions are that cancer is a 'modern disease' and that it primarily afflicts humans. Nothing could be farther from the truth. Cancer or cancer-like phenomena are found in almost all mammals, birds, reptiles, insects and even plants. Work by Athena Aktipis and her collaborators shows the existence of cancer, or cancer analogues, across all metazoan categories, including fungi and corals.[29] (See Fig. 13.) Instances of cancer have even been found in the simple organism hydra.[30]

The fact that cancer is so widespread among species points to an ancient evolutionary origin. The common ancestor of, say, humans and flies dates back 600 million years, while the broader categories of cancer-susceptible organisms have points of convergence over 1 billion years ago. The implication is that cancer has been around for as long as there have been multicelled organisms (metazoa). This is reasonable. It goes without saying that cancer is a disease of bodies; it makes little sense to say that an isolated bacterium has cancer. But bodies did not always exist. For 2 billion years life on Earth consisted of single-celled organisms only. About 1.5 billion years ago the first multicellular forms appeared, during the geological epoch known as the Proterozoic ('earlier life' in Greek).*

The transition to multicellularity entailed a fundamental change in the logic of life. In the world of single cells, there is but one imperative: replicate, replicate, replicate! In that sense, single cells are immortal. Multicelled creatures, however, do things very differently. Immortality is outsourced to specialized germ cells (for example, eggs and sperm) whose job is to carry the organism's genes forward into future generations. Meanwhile, bodies, which are a vehicle for these germ cells, behave very differently. They are mortal. The cells of the body (somatic cells) retain a faint echo of their past immortality in a limited ability to replicate. A typical skin cell, for example, can divide – between fifty and seventy times. When a given somatic

---

* Multicellularity arose independently several times. True multicellularity is restricted to eukaryotes. However, bacteria can aggregate into colonies that sometimes display cancer-like phenomena.

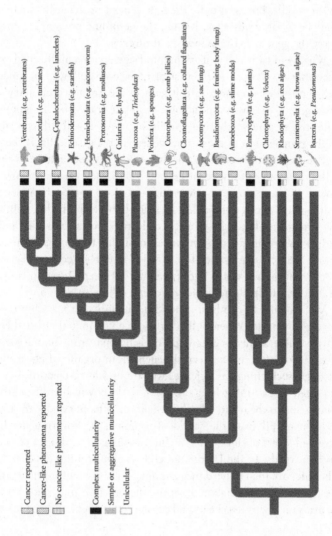

Fig. 13. Cancer across the tree of life.

Vertebrata (e.g. vertebrates)
Urochordata (e.g. tunicates)
Cephalochordata (e.g. lancelets)
Echinodermata (e.g. starfish)
Hemichordata (e.g. acorn worm)
Protosomia (e.g. molluscs)
Cnidaria (e.g. hydra)
Placozoa (e.g. *Trichoplax*)
Porifera (e.g. sponges)
Ctenophora (e.g. comb jellies)
Choanoflagellata (e.g. collared flagellates)
Ascomycota (e.g. sac fungi)
Basidiomycota (e.g. fruiting body fungi)
Amoebozoa (e.g. slime molds)
Embryophyta (e.g. plants)
Chlorophyta (e.g. *Volvox*)
Rhodophyta (e.g. red algae)
Stramenopila (e.g. brown algae)
Bacteria (e.g. *Pseudomonas*)

Cancer reported
Cancer-like phenomena reported
No cancer-like phenomena reported

Complex multicellularity
Simple or aggregative multicellularity
Unicellular

cell reaches its use-by date* it either goes dormant (a state called *senescence*) or commits suicide (*apoptosis*). That does not spell the end of the organ, because replacement cells of the same type are made by stem cells. But eventually the replacement process also wears out and the whole body dies, leaving the germ-line progeny, if any, to carry the genetic heritage into the future.

Why would any cell in its right mind sign up for a multicelled existence that involves a short burst of replication followed by suicide? What possible advantage can it gain in the great evolutionary survival game? As always in biology, there are trade-offs. By joining a collective of genetically similar cells, a given cell will still contribute to the propagation of most of its genes via the germ cells. If the collective as a whole possesses survival functions unavailable to single cells, then the arithmetic of genetic legacy may tip the balance in favour of the community over the go-it-alone approach. When the mathematics looks right, a deal is struck between individual cells and the organism. The cells join the collective project and die, and in return the organism takes on the responsibility to propagate the cells' genes. Multicellularity therefore involves an implicit contract between the organism as a whole and its cellular members. It was a contract first signed in the Proterozoic era, over a billion years ago.

Multicellularity can be a good idea – it works for us! – but it does have a downside. When individuals join a communal effort there is always vulnerability to cheating. This is familiar from human society, where people receive a survival benefit from organized government through such things as defence, welfare and infrastructure but are expected to pay for it in taxes. As is well known, there is a strong temptation to cheat – to take the benefits on offer but dodge the taxes. It happens all over the world. To counter it, governments have invented layers of rules. (Tax law in Australia, for example, runs to a million words. In the US the tax code is almost infinitely complex.) The rules are then policed by government and law-enforcement agencies. In spite of this elaborate set-up, the system is imperfect; there is an arms race between cheats and enforcers: internet fraud and identity

---

* When is that? When little caps on the end of chromosomes, called telomeres, are worn down.

theft are prime current examples. A similar arms race is played out in multicellularity. To get individual cells to stick to the contract there have to be layers of regulatory control, policed by the organism as a whole, to deter cheats. Thus, a given somatic cell (skin cell, liver cell, lung cell . . .) will normally divide only when the regulations permit. When more cells of that variety are needed, the cell's own internal 'replication program' will take care of it. But if division is inappropriate, then the regulatory mechanisms will intercede to either prevent it or, if the cell is persistently recalcitrant, order the death sentence: apoptosis. A graphic example of this strict policing occurs if a cell finds itself in the wrong tissue environment. For example, if a liver cell is accidentally transported to, or deliberately transplanted in, the lung, it won't fare well. Chemical signals from the lung tissue recognize that they have an interloper ('not one of us!') and may order apoptosis.

What does cheating mean for a cell in a multicelled organism? It means a default to the selfish every-cell-for-itself strategy of unicellular life: replicate, replicate, replicate. In other words, uncontrolled proliferation. Cancer. Put simply, cancer is a breakdown of the ancient contract between somatic cells and organisms, followed by a reversion to a more primitive, selfish agenda.

Why does the policing fail? There can be many reasons. An obvious one is damage, say from radiation or a carcinogenic chemical, to one of the 'police genes'. There is a class of genes, of which p53 is the best known, that serve as tumour suppressors. Damage p53 and a tumour may not be suppressed. Another trigger is immunosuppression. The adaptive immune system includes cancer surveillance as part of its remit. If the system is working properly, incipient cancer cells are spotted and zapped (or incarcerated and contained) before they can cause trouble. But cancer cells can cloak themselves chemically to hide from the immune surveillance police. They can also subvert the immune system by recruiting its scouts (macrophages) and 'turning' them, like captured spies, to work for them. Tumour-associated macrophages will screen a tumour and stymie the immune attack.

For cancer to take hold, two things have to happen. A normal cell has to embark on a cheating strategy and the organism's police have to slip up somewhere. The conventional explanation is the somatic mutation theory, according to which genetic damage accumulates in somatic cells as a result of ageing, radiation or carcinogenic chemicals causing the cells to misbehave and go rogue, that is, embark on their own agenda. The resulting 'neoplasm', or population of new cells, rapidly develops (says the orthodox theory) the distinctive hallmarks I mentioned, including uncontrolled proliferation, plus a tendency to spread around the body and colonize remote organs. The somatic mutation theory assumes that the *same* hallmarks of cancer are reinvented *de novo* in each host solely by a sort of fast-paced Darwinian process of natural selection, in which the fittest (that is, nastiest) cancer cells outbreed their competitors via runaway replication, eventually killing the host (and themselves). Though entrenched, the somatic mutation theory has poor predictive power, its explanations amounting to little more than *Just So* stories on a case-by-case basis. Most seriously, it fails to explain how mutations confer so many fitness-improving gains of function in a single neoplasm in such a short time (yes, that again). It also seems paradoxical that increasingly damaged and defective genomes should enable a neoplasm to acquire such powerful new functionality and so many predictable hallmarks.

TRACING THE DEEP EVOLUTIONARY ROOTS OF CANCER

Over the past few years my colleagues and I have developed a somewhat different explanation of cancer that seeks its origins in the far past.[31] We were struck by the fact that cancer almost never invents anything new. Instead, it merely appropriates already existing functions of the host organism, many of them very basic and ancient. Limitless proliferation, for example, has been a fundamental feature of unicellular life for aeons. After all, life is in the business of replication and cells have had billions of years to learn how to keep going in the face of all manner of threats and insults. Metastasis – the process whereby normally sedentary cells become mobile, quitting a tumour to spread around the body – mimics what happens during early-stage embryogenesis, when immature cells are often not anchored in place but surge in organized patterns to designated locations. And the propensity of circulating cancer cells to invade other organs closely parallels what the immune system does to heal wounds. These facts, which oncologists know well, combined with the predictable and efficient way that cancer progresses through its various stages of malignancy, convinced us that cancer is not a case of damaged cells randomly running amok but an ancient, well-organized and efficient survival response to stress.* Crucially, we believe that the various distinctive hallmarks of cancer do not independently evolve as the neoplasm goes along – that is, are stumbled across by accident – but are deliberately switched on and deployed systematically as part of the neoplasm's organized response strategy.

In summary, our view of cancer is that it is not a *product* of damage but a systematic *response* to a damaging environment – a primitive cellular defence mechanism. Cancer is a cell's way of coping with a bad place. It may be *triggered* by mutations, but its root cause is the self-activation of a very old and deeply embedded toolkit of emergency survival procedures.† The key distinction between the two theories can be illustrated with an analogy. Consider a playground victim of bullying who runs away as a survival strategy. The victim's exit is self-propelled; the pushes and punches of his attackers may *trigger* his flight, but they are not themselves the ultimate *cause* of his motion – he is not pushed away, he *runs* away. Here is another analogy. If a computer suffers an insult – corrupted software or a mechanical blow – it may start up in safe mode (see Fig. 14). This is a default program enabling the computer to run on its core functionality even with the damage. In the same way, cancer is a default state in which a cell under threat runs on *its* ancient core functionality, thereby preserving its vital functions, of which proliferation is the most ancient, most vital and most protected. To trigger cancer, the threats don't have to be radiation or chemicals; they could be ageing tissues, low-oxygen

* The use of the term 'stress' here, as previously, refers to a threatening micro-environment, e.g. carcinogens, radiation or hypoxia. The widespread belief that people who *feel* stressed may get cancer is not obviously related to the physical stress I am discussing here.

† The distinction between trigger and root cause is analogous to running a basic and well-used computer software package, e.g. Microsoft Word. The 'open' command triggers Word, but the 'cause' of the 'Word phenomenon' is the Word software, which has its origin in the dim and distant past of the computer industry.

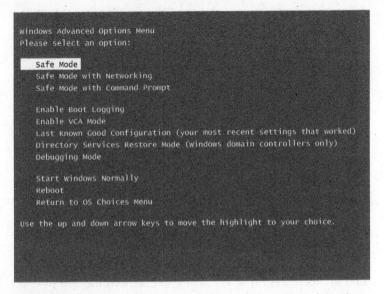

Windows Advanced Options Menu
Please select an option:

    Safe Mode
    Safe Mode with Networking
    Safe Mode with Command Prompt

    Enable Boot Logging
    Enable VGA Mode
    Last Known Good Configuration (your most recent settings that worked)
    Directory Services Restore Mode (windows domain controllers only)
    Debugging Mode

    Start Windows Normally
    Reboot
    Return to OS Choices Menu

Use the up and down arrow keys to move the highlight to your choice.

Fig. 14. This depressing screen may appear on your computer when it has a problem booting up. It indicates damage of some sort, causing the operating system to run on its core functionality while the problem is addressed. Cancer could be doing something similar – defaulting to the cell's core functionality, which evolved more than a billion years ago, while ignoring or disabling the more recently evolved biological 'bells and whistles'.

tension or mechanical stresses of various sorts, including wounding. (Or even electrical disruption – see p. 114.) Many factors, individually or collectively, can cause the cell to adopt its inbuilt 'cancer safe mode'.

Although elements of the cancer default program are *very* ancient, dating back to the origin of life itself, some of the more sophisticated features recapitulate later stages in evolution, especially in the period between 1.5 billion and 600 million years ago, when primitive metazoans emerged. In our view, cancer is a sort of throwback or default to an ancestral form; in technical jargon, it is an *atavistic* phenotype. Because cancer is deeply integrated into the logic of multicellular life, its ancient mechanisms highly conserved and fiercely protected, combating it proves a formidable challenge.

Our theory makes many specific predictions. For example, we expect the genes that are causally implicated in cancer (usually called oncogenes) to cluster in age around the onset of multicellularity. Is there any evidence for this? Yes, there is. It is possible to estimate the ages of genes from their sequence data by comparing the number of differences across many species. This well-tried technique, known as phylostratigraphy, enables scientists to reconstruct the tree of life, working backwards from common features today to deduce the convergence point in the past (see Fig. 15).

A study in Germany using four different cancer gene datasets demonstrated the presence of a marked peak in genes originating at around the time that metazoa evolved.[32] A recent analysis of seven tumour types led by David Goode and Anna Trigos in Melbourne, Australia, focused on gene expression.[33] They sorted genes into sixteen groups

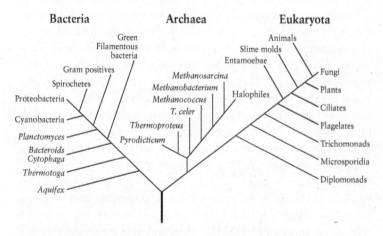

Fig. 15. Tracing the history in the tree of life. Since Darwin first drew a tree doodle to represent the divergence of species over time, biologists have attempted to reconstruct the history of life using the fossil record. Now they also use a method called phylostratigraphy, which appeals to gene sequences across many species to determine common ancestors in the far past. The tree above shows the three great domains of life diverging from a common ancient origin. The lengths of the lines indicate genetic distance. The last common ancestor of this tree lived about 3.5 billion years ago.

by age and then compared expression levels in cancer and normal tissue for each group. The results were striking. Cancer over-expresses genes belonging to the two older groups and under-expresses younger genes, exactly as we predicted. Furthermore, they found that as cancer progresses to a more aggressive, dangerous stage the older genes are expressed at higher levels, confirming our view that cancer reverses the evolutionary arrow at high speed as it develops in the host organism, with the cells reverting to their primitive ancestral forms in a space of weeks or months. More generally, the Australian group found that genes associated with unicellularity are more active in cancer than those that evolved later, during the era of multicellularity.

In our own work at Arizona State University we looked at mutation *rates*.[34] The atavistic theory predicts that older genes should be less mutated in cancer (after all, they are responsible for running the 'safe mode' program), while younger genes should be mutated more. My colleagues Kimberly Bussey and Luis Cisneros considered a total of 19,756 human genes and used an inventory of cancer genes called COSMIC compiled by the UK Sanger Institute. This data was combined with a database of genetic sequence data from about 18,000 species across all taxonomic groups that includes an analysis of evolutionary ages. This allowed my colleagues to estimate the evolutionary ages of the genes in the human genome. They found that genes younger than about 500 million years were indeed more likely to be mutated – in normal, but especially in cancer, tissue – while genes older than a billion years tend to suffer less mutations than average, as we expected. They also confirmed the German study that the ages of cancer genes* display a cluster around the time of the onset of multicellularity, supporting our contention that cancer is driven by disruption of functions that evolved to achieve multicellular organization. COSMIC classifies genes into dominant and recessive, and my colleagues found that cancer genes with recessive mutations were significantly older than most human genes.

The most telling result came from addressing a rather different

---

* Defined in this study to be those genes demonstrated to be causally implicated in cancer.

question: what are cancer genes good for? A database called DAVID organizes genes around their functionality. When Cisneros and Bussey fed the COSMIC data into DAVID what leapt out was that recessive cancer genes older than 950 million years were strongly enriched for two core functions: cell cycle control and DNA damage repair involving double-strand breaks (the worst kind of damage DNA can suffer). Looking at the evolutionary history of the genes involved, the researchers spotted something significant. The non-mutated genes in the same DNA repair pathways correspond to* those genes in *bacteria* that drive the adaptive mutation response to stress – the very phenomenon I discussed earlier in this chapter (see p. 124). And as in bacteria, these genes serve to turn up the cell's mutation rate in a desperate effort to survive by evolving a pathway out of trouble. I explained Susan Rosenberg's discovery that when a bacterium senses a double-strand break it can switch to a sloppy repair mechanism, creating a trail of errors (mutations) either side of the break. We were keen to know whether cancer cells also display a pattern of damage around double-strand repairs. It turns out they do. Out of 764 tumour samples from seven different sites (pancreas, prostate, bone, ovary, skin, blood and brain) my colleagues looked at, 668 had evidence of mutational clustering around the break. This all fitted in with our theory that cells under stress turn cancerous by reawakening ancestral gene networks which, among other things, create a high rate of mutations. Thus, one of the best-known hallmarks of cancer (and the main reason why it so often evades chemotherapy by evolving drug-resistant variants) turns out to be self-inflicted. This finding fits in well with the atavism theory: cancer is merely appropriating an ancient stress response, still used by bacteria today, which evolved way back in the era of unicellular life.

As in stressed bacteria, mutations in cancer cells are far from random: there are definite mutational 'hotspots' and 'cold-spots' (regions of low mutation). This makes perfect sense. Multicelled organisms should work hard to protect key parts of their genomes, such as those responsible for running the core functions of the cell, and devote less resources

---

* Technically, they are termed 'orthologs' of those genes.

to the 'bells and whistles' associated with more recently evolved and less critical traits. A project led by Princeton's Robert Austin and Amy Wu subjected cancer cells to a therapeutic toxin (doxorubicin) and studied the evolution of their resistance to this drug. Austin and Wu found that cold-spot genes were significantly older than average.[35] These new results help explain why natural selection hasn't eliminated the scourge of cancer. If tumours really are a reversion to an ancestral form, then we might expect that the ancient pathways and mechanisms that drive cancer would be among the most deeply protected and conserved, as they fulfil the most basic functions of life. They can't be got rid of without disaster befalling the cells concerned. The mutator genes we investigated are just one example.

Another reason that evolution hasn't eliminated cancer is because of the link with embryogenesis. It has been known for thirty years that some oncogenes play a crucial role in development; eliminating them would be catastrophic. Normally, these developmental genes are silenced in the adult form, but if something reawakens them cancer results – an embryo gone wrong developing in adult tissue. The writer George Johnson summarizes this well by referring to tumours as the 'embryo's evil twin'.[36] Significantly, the early stages of an embryo are when the organism's basic body plan is laid down, representing the earliest phase of multicelled life. When the cancer switch is flipped, there will be systematic disruption in both the genetic and epigenetic regulators of information flow, as the cells recapitulate the very different circumstances of early embryo development. This will involve both changes to the way regulatory genes are wired together and changes in patterns of gene expression. Our research group is trying to find information signatures of these changes. We hope it will prove possible to identify distinct 'informational hallmarks' of cancer to go alongside the physical hallmarks I mentioned – a software indicator of cancer initiation that may precede the clinically noticeable changes in cell and tissue morphology, thus providing an early warning of trouble ahead.

The atavistic theory of cancer has important implications not just for diagnosis but for therapy. We think the search for a general-purpose 'cure' for cancer is an expensive diversion, and that cancer, being so deeply entrenched in the nature of multicellular life itself, is best managed and controlled (not eliminated) by challenging the cancer with physical conditions inimical to its ancient atavistic lifestyle. Only by fully understanding the place of cancer in the overall context of evolutionary history will a serious impact be made on human life expectancy in the face of this killer disease.

5

Spooky Life and Quantum Demons

It is often said that however ingenious mankind's inventions are, nature invariably beats us to it. And it's certainly true that biology discovered wheels and pumps, scissors and ratchets, long before we did. But that's not all. Nature also discovered digital information processing billions of years before humans invented the computer. Today, we are on the threshold of a new technological revolution, and one that promises changes as sweeping as those that followed the advent of the digital computer. I am referring to the long-sought-after quantum computer.

The essential idea was captured by Richard Feynman in a futuristic lecture entitled 'Simulating Physics with Computers' delivered at the University of California, Berkeley, in 1982.[1] Feynman pointed out that when conventional computers are used to model fundamentally quantum mechanical objects such as molecules, they struggle with the sheer computational resources needed to keep track of everything. He conjectured, however, that a (then hypothetical) quantum computer would be up to the job, because it would be simulating something of its own basic kind. In 1985 Oxford physicist David Deutsch took the idea further, working out the precise rules whereby information could be inscribed in the states of atomic and subatomic systems and then manipulated using the standard laws of quantum mechanics.

The secret of a quantum computer is something called superposition. In a conventional (classical) computer a switch is definitely either on or off, representing 1 or 0. In a quantum computer it can be *both*, so it can represent 1 and 0 at the same time – a 'superposition' of 1 and 0. The superposition is not merely a fifty–fifty mix of the two numbers but all possible blends. Physicists refer to such an entity as a *qubit* (for 'quantum bit'). Tangle just a few dozen qubits together and you can, in principle, create a device that would outperform the best conventional computer.

It wasn't long before physicists started scrambling to make such a thing, and today the race to develop and market a fully functional quantum computer is the primary goal of a multibillion-dollar industry involving major government and commercial research programmes around the world. The huge investment is because of the sheer computational power a quantum computer would unleash. Not only would it be able to simulate atomic and molecular processes in detail, it could crack the codes used to encrypt most of our communications and sort through vast databases at lightning speed. If it were made widely available, conventional computer security would be a joke: a quantum computer would jeopardize the intelligence services, diplomatic communications, banking transactions, internet purchases – in fact, anything online that is confidential and requires encryption.

But what has quantum computation got to do with life? Well, in biology, the name of the game is information management. Given that life is so adept at manipulating bits, might it have learned to manipulate qubits too? A number of such claims have been made. Though there is scant evidence that organisms engage in actual quantum computing, it is becoming increasingly clear that life does indeed harness some quantum effects.

QUANTUM THEORY IS
SERIOUSLY WEIRD

Einstein once described quantum effects as 'spooky'. In fact, he thought they were so spooky he steadfastly maintained that quantum mechanics gave a flawed account of nature. It seemed in conflict with his cherished theory of relativity by permitting faster-than-light effects* and he was uncomfortable with the idea that uncertainty and indeterminism underpinned fundamental phenomena. Today, very few physicists would side with Einstein. Quantum mechanics, with all its spookiness, has embedded itself firmly in mainstream physics. After all, here is a theory that not only explains almost everything from subatomic particles to

* We now know that quantum mechanics does not permit faster-than-light communication, undercutting Einstein's objection.

stars, it has given us indispensable forms of technology such as the laser, the transistor and the superconductor. The problem is not with the extraordinary power of quantum mechanics to explain the world and to drive the growth industries of the twenty-first century. It lies with its seriously weird implications for the nature of reality.

Imagine the following scenarios:

You throw a tennis ball at a window pane and it bounces back. You throw it again, exactly as before, and the ball appears on the far side of the glass without breaking it.

You strike a billiard ball directly towards a pocket. When it reaches the edge of the hole, instead of dropping in it bounces right back at you as if the hole's near edge were an invisible wall.

A ball is rolling along a gutter in the road towards an intersection. When it gets there it turns the corner on its own, without any need for a sideways kick.

These events would be regarded as miraculous if they occurred in daily life, but they happen all the time at the atomic and molecular level, the domain of quantum physics. Other weird quantum effects with no everyday counterpart include particles such as electrons seemingly being in two places at once,* a pair of photons metres apart spontaneously coordinating their activities (what Einstein called 'spooky action-at-a-distance'), and a molecule spinning both clockwise and anticlockwise at the same time. Entire books have been written about these peculiar yet real effects. Here I am concerned with one question only: do spooky quantum effects occur in biology?

There is a trivial sense in which all biology is quantum. Life is, after all, applied chemistry, and the shapes, sizes and interactions of molecules all need quantum mechanics to explain them. But that's not what people have in mind when they talk about 'quantum biology'. What we really want to know is whether *non-trivial* quantum processes such as tunnelling (ball-through-the-window-pane effect) or what is referred to as entanglement (spooky action-at-a-distance) have an important role to play in life.

---

* This oft-repeated description must be interpreted cautiously. The electron doesn't literally bifurcate. Any experiment to determine its precise location will always find it in one place *or* the other. But in the absence of such a measurement, there are definite physical effects stemming from its indeterminate position.

Box 11: Is it a wave? Is it a particle? No, it's a quantum!

One of the founding discoveries of quantum physics is that waves can sometimes behave like particles. Einstein was the first to suggest this, in 1905, with his hypothesis that light, known to be a wave, trades energy only in discrete packets, called photons. Conversely, electrons, which we normally think of as particles, sometimes behave like waves (as do all particles of matter). It was Schrödinger who wrote down the equation that describes how matter waves behave. This 'wave-particle' duality is at the heart of quantum mechanics. Many surprising quantum effects, such as the three ball scenarios I gave above, are consequences of the wave nature of matter. It is futile to argue about whether a photon or an electron is 'really' a wave or a particle; experiments can be done in which either of these aspects is manifested, though never both together. One simply has to accept that the denizens of the microworld have no everyday counterpart.

The way quantum waves spread and merge critically affects their physical significance. Imagine dropping two stones together into a smooth pond, a few feet apart. Ripples from each stone spread out and overlap. Where the peak of a wave from one stone meets the peak of the wave from the other, they reinforce to make a higher peak. Where peak meets trough, they cancel. The resulting criss-cross pattern of the merged waves, if nothing disturbs it, is called *coherent*. But now suppose there is a hailstorm spattering the pond with many additional ripples. The nice orderly wave pattern from the stones will be disrupted. This is called *decoherence*. Coming to quantum matter waves, many weird quantum effects stem from the coherence of the waves. If coherence disappears, so do most of the non-trivial quantum effects. Like water waves, electron waves will decohere if they are disturbed. Electrons in living matter don't suffer from hailstorms, but they do have to contend with molecular storms, such as the incessant thermal bombardment by water molecules. Back-of-the-envelope calculations suggest that decoherence will occur extremely rapidly under most biological conditions. But there seem to be escape clauses that permit anomalously slow decoherence rates under some special circumstances.

As a rule, if something gives life an edge, even a slight one, then natural selection will exploit it. If 'something quantum' can enable life to go faster, cheaper, better, we might expect evolution to stumble across it and select it. Right away, however, we hit a snag with this glib reasoning. Quantum effects represent a subtle and delicate form of atomic and molecular order. The enemy of all quantum effects is *disorder*. But life is awash with disorder! That inescapable clamour of randomly agitated molecules, the pervasive depredations of the second law of thermodynamics. Entropy, entropy everywhere! Non-trivial quantum effects can survive in the face of this inexorable thermal noise only under very peculiar circumstances. Visit any laboratory that studies quantum phenomena. You will see gleaming steel chambers, winking electronics, humming cryostats, multitudinous wires and pipes, meticulously aligned laser beams, computers – a plethora of expensive, high-precision, finely tuned equipment. The primary purpose of all that fancy gadgetry is to reduce the disturbance wrought by thermal agitation, either by screening it out – isolating the quantum system of interest from the surroundings – or cooling everything relevant to near absolute zero (about −273°C). A big slice of the fortune pouring into the quantum computer industry is directed to combating the ever-present menace of thermal disruption. It is proving to be very, very hard.

In view of the extraordinary lengths to which physicists have to go in order to evade the effects of thermal noise, it seems incredible that anything spooky would be going on in the messy, relatively high-temperature world of biological organisms. A protein in a cell is about as far from an isolated low-temperature system as one can imagine. But remember the demon. It was designed by Maxwell precisely to conjure order out of chaos, to cheat the second law, to evade the corrosive effects of entropy. Although we know that even a demon can't violate the letter of the second law, it can certainly violate the spirit of it. And life is replete with demons. Could it be that, among their ingenious repertoire of tricks, life's demons have also learned how to juggle, not just bits, but qubits too, with a dexterity as yet unmatched by our state-of-the-art laboratories?

TUNNELLING

Stuart Lindsay, a colleague of mine at ASU, is a real-life quantum biologist. The focus of his research is the investigation of how electrons flow through organic molecules, especially the As, Cs, Gs and Ts of DNA fame. The method they have perfected in his lab is to unzip the DNA double helix into single strands and suck one of them, spaghetti-like, through a tiny hole – a 'nanopore' – in a plate. Positioned across the hole is a pair of miniature electrodes. As each 'letter' transits the hole, electrons go through it, creating a tiny current. Happily, it turns out that the strength and characteristics of the current differs discernibly for each letter, so Lindsay's set-up can be used as a high-speed sequencing device. He has also found that amino acids can be good electrical conductors too, opening the way to sequencing proteins directly.

When Lindsay first described his work, I confess I was puzzled by why organic molecules would conduct any electricity at all. After all, we use organic substances like rubber and plastic as insulators, that is, as a *barrier* to electricity. And in fact, at first sight it's hard to see how electrons would find a path through, say, a nucleotide or an amino acid. The explanation, it turns out, lies with a curious quantum phenomenon known as tunnelling – the 'ball-through-the-window-pane' effect. Electrons can traverse a barrier even when they have insufficient energy to surmount it; if it weren't for the wave nature of matter (see Box 11), the electrons would just bounce right back off the organic molecule. The tunnel effect was predicted when Schrödinger presented his famous equation for matter waves in the 1920s and examples were soon found. A type of radioactivity known as alpha decay, first observed in the 1890s, would be incomprehensible were it not for the fact that the emitted alpha particles tunnel their way through the nuclear force barrier of uranium and other radioactive substances. Electron tunnelling forms the basis of many commercial applications in electronics and materials science, including the important scanning tunnelling electron microscope.

Stuart Lindsay can send electrons through organic molecules, but does nature do it too? It does indeed. There is a class of molecules known as metallo-proteins, basically proteins with a metal atom such as iron entombed within (a well-known example is haemoglobin). Metals are good conductors, so that helps. But the phenomenon of tunnelling through organic molecules is actually quite widespread. Which raises a curious question: why do electrons want to traverse proteins anyway? One reason it's a good thing is metabolism. Enzymes connected with oxygenation, and the synthesis of the all-important energy molecule ATP, hinge on rapid electron transport. Slick electron tunnelling greases the wheels of life's energy-generating machine. This is not just a happy coincidence; these organic molecules have been honed by evolution. Any old jumble of organic molecules won't do, at least according to Harry Gray and Jay Winkler at Caltech's Beckman Institute: 'Stringent design requirements must be met to transport charges rapidly and efficiently along specific pathways and prevent the off-path diffusion . . . and the disruption of energy flow.'[2]

While all this is very interesting physics, there is a fascinating bigger question. Have biomolecules more generally been selected by evolution for efficient quantum 'tunnellability'? A recent analysis by Gábor Vattay of the Eötvös Loránd University in Hungary and his collaborators suggests that 'quantum design' may not be restricted to metabolism but is a generic feature of biology.[3] They arrive at this conclusion by studying where on the spectrum between electrical conductors and insulators key biological molecules lie. They claim to have identified a new class of conductors that occupies the critical transition point between the material behaving like an insulator and a disordered metal; many important biomolecules apparently fall into this category. Testosterone, progesterone, sucrose, vitamin D_3 and caffeine are among many examples cited by Vattay et al. In fact, they believe that 'Most of the molecules taking part actively in biochemical processes are tuned exactly to the transition point and are critical conductors.'[4] Being poised at the edge of the ability to conduct electricity is likely to be a rather rare property of a molecule, and given the astronomical number of possible molecules that could be formed from the building blocks that life uses, the chances of hitting an arrangement that confers such critical conductance is infinitesimal.

Hence there must have been strong evolutionary pressure at work. In this case at least, it looks like biology has indeed spotted a quantum advantage and gone for it.

TRIPPING THE LIGHT FANTASTIC

Quantum biology lay largely in the shadows until 2007, when a dramatic discovery cast light on it – quite literally – and propelled the subject to world attention. A group of scientists at the University of Chicago led by Greg Engel was investigating the physics of photosynthesis.[5] Now you might think that photosynthesis is by definition a quantum phenomenon – after all, it involves photons. But that merely puts it into the category of 'trivial' quantum effects. The organism – it could be a plant or a photosynthetic bacterium – uses light to make biomass from carbon dioxide and water. In that respect, the photon is simply a source of energy; its quantum aspect is incidental. Where the spooky stuff starts is at the next step. The molecular complex that captures the photon and the reaction centre where the actual chemistry is done are not the same. It's rather like having solar panels in a field to power a factory located down the road. In biology, there's always a competition for energy, so it pays to avoid wasting too much of this valuable resource when passing it from place to place, in this case from the light-harvesting molecules to the reaction centre. Scientists have long been mystified about how photosynthesis can accomplish this transfer so efficiently. Now it seems that non-trivial quantum effects could pave the way.

To explain what's going on I need to invoke another weird quantum property already mentioned in passing: the ability of quantum particles to be 'in two places at once'. In fact, they can be in many places at once. A corollary of this is that in going from A to B a particle can take more than one route simultaneously. To be precise, it takes *all possible* routes, and not just the shortest one (see Fig. 16). The strange calculus of quantum mechanics requires one to integrate *all* available pathways between start and finish: they all contribute to how the particle gets there. This sounds totally mysterious, but it's not if the particle is viewed as a wave which spreads out rather than as a little blob which doesn't. Think of a water wave approaching a bollard sticking out from the sea bed. The waves curve around it, some going left, some going right, and they join up on the other side. Quantum waves do the same. Imagination fails us, however, when we try to think in terms of particles: how can a *single* particle go everywhere at once? How can one envisage that? A popular interpretation of 'what is going on' in quantum mechanics is to think (in this example) of each pathway from A to B as representing a separate world. If there is an obstacle in the path of the particle (analogous to the bollard in the water), well, in some worlds the particle goes to the right and in others it goes to the left.

Of course, people ask, 'But which way did it go really?' The answer depends on what you mean by 'really', which is where discussions of

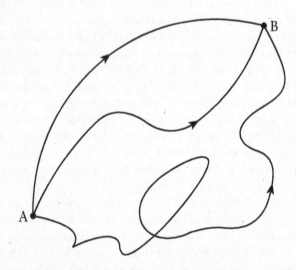

Fig. 16. Quantum paths. In daily life, if a particle (e.g. a cricket ball) travels from point A to point B, it follows a definite path in space between them. Not so for atoms and subatomic particles. According to the weird rules of quantum mechanics, a particle takes *all possible paths* between A and B in a ghostly amalgam. Every path contributes to the properties of the particle; they do have a real effect.

quantum mechanics start to become murky and many people are left behind. Nevertheless, I shall endeavour to explain it. In the old-fashioned approach (going back decades), these alternative worlds (each world containing just one particle trajectory) were considered merely *contenders* for reality, ghostly virtual worlds that don't 'really exist' but collectively form an amalgam – a superposition – from which 'the real world' of experience emerges. For definiteness, consider that an experimenter dispatches an electron from a well-defined point A and detects it at a well-defined point B; well, according to quantum mechanics, it is not possible to say *how* it got from A to B. There is no 'fact of the matter' about the intervening route. Not only is it impossible for the experimenter to know the route, even *nature* doesn't know. If you try to station a detector partway between A and B to sneak a look, it totally changes the whole result. The physicist John Wheeler, a doyen of colourful descriptions, liked to say that quantum propagation (between A and B, as I have described it) is 'like a great smoky dragon'. It has 'sharp teeth' and a 'sharp tail' (at A and B, where the experimenter receives sharply defined information on the particle's whereabouts) but in between all is veiled in smoke.

These days, many leading physicists insist that the multiplicity of different quantum worlds are in fact *real* worlds; they exist in parallel, a point of view known as the many-worlds, or many-universes, interpretation of quantum mechanics. As to why we experience only one world, one has to ask what is meant by 'we'. Suppose each world has a separate version of you. There are now many worlds and many (nearly identical) yous. Each version of you sees just one world. Whether one (ones?) buys into this fashionable but extravagant interpretation of quantum mechanics doesn't matter here: one can at least safely say that in going from A to B the particle can *try out all routes together*.

Does the route fuzziness matter, or is this all philosophical mumbo-jumbo? It certainly does matter, because the alternative paths interfere with each other (like the water waves going around the bollard). Sometimes this interference creates 'no-go zones' for the particle (where two merging waves cancel out, peak to trough); conversely, it may facilitate its appearance in another region (where the waves reinforce).

Quantum interference effects of just this nature seem to make a difference in the molecular complex responsible for photosynthesis. The Chicago team, soon joined by Graham Fleming and his group at UC Berkeley, focused their attention on green sulphur bacteria. These inconspicuous microbes live in lakes and around deep ocean volcanic vents as far down as 5 kilometres. No sunlight penetrates to that depth, but the hot vents emit a dim red glow and it is from this feeble light source that the bacteria make a living. 'Feeble' is the word: it's been estimated that each photosynthetic complex gets only about one photon a day. That's a trillion-trillionth of what a plant leaf might expect. With so few photons to go round, green sulphur bacteria need to do the very best with what they can get and, indeed, efficiencies approach 100 per cent, with little or no energy squandered.

Here is how it works. The photons come in, one by one, and are absorbed somewhere within a bundle of light-harvesting antennae, each packed with a type of chlorophyll (200,000 molecules in total). About a picosecond (one trillionth of a second) later the captured energy appears in the chemical reaction centre. To get there, the energy traverses what may be loosely compared to the cable or waveguide that connects the antenna on your roof to your television. (At least it did in the days before optical cable TV.) In the photosynthesis case, the role of the cable is played by a molecular bridge called an FMO complex which is made up of eight molecular subunits 1.5 nanometres apart, each of which is also made of chlorophyll, fixed to a protein scaffold. The photon itself is absorbed and disappears, but its energy is captured (in a form I shall describe shortly) and enters the FMO complex through a molecular structure that biologists engagingly call a 'baseplate', where it is received by one of the FMO subunits, and is then passed among the rest like a relay baton, until it reaches a subunit adjacent to the all-important 'factory' – the reaction centre – where it is handed over to power the chemical reaction. The whole process is a race against time, to deliver the goods before some external disturbance disrupts it.

The whole set-up may seem a bit complicated and ramshackle, with lots of scope for error, delay and 'dropping the baton' on the way. All the molecules involved are big and complicated and they are jiggling about because of thermal agitation; it's easy to imagine the precious energy, painstakingly harvested and destined as it is for the reaction centre, ending up instead being scattered and dissipated into the messy intervening infrastructure. But that doesn't happen. The energy arrives unmolested and in record time. It used to be supposed that this energy transport traversed the FMO complex in a series of simple hops (or baton passes), haphazardly accomplished amid the thermal clamour of the molecular milieu. But that looks too hit and miss for such a fine-tuned mechanism. Which is where quantum mechanics comes in.

To give the gist of the quantum explanation, first let me explain in what form this energy is stored. When the photon is absorbed it releases an electron from the antenna molecule (this is the familiar photoelectric effect), leaving behind a positively charged 'hole'. Because the electron is embedded in a molecular matrix, it doesn't fly off, free. Instead, it remains loosely bound to the hole in a very large orbit (physicists say it is 'delocalized'); the arrangement is called an 'exciton'. The exciton can itself behave in many respects like a quantum particle, with associated wavelike properties, and it is this exciton, not an electron as such, that is passed through the FMO complex. Viewed in terms of pathways, there are many routes the exciton can take and, if quantum coherence is maintained, *will* take – simultaneously. Loosely speaking, the exciton is able to sift all the options at once and feel out the best possible route to the reaction centre. And then take it. What I am describing is an extraordinary type of demon, a quantum super-demon that 'knows' all available pathways at once and can pick the winning one. In more careful physics-speak, the claim is that constructive interference occurs across multiple molecules in the FMO complex so that coherent excitons can optimize the efficiency and deliver the energy to the reaction centre before it can be dissipated into the molecular environment. It takes about 300 femtoseconds (a femtosecond is a thousandth of a trillionth of a second) to get there.

To study this complicated mechanism the Berkeley group used ultra-fast lasers to excite an FMO complex in the lab. They were able to follow the fortunes of the energy as it sashayed through the molecular maelstrom and announced that some sort of 'quantum beating' effect – coherent oscillations – did indeed contribute to the high-speed transfer of energy.

The results of these experiments came as a complete surprise because it seemed that the excitons' carefully balanced dance would be wrecked by thermal agitation. At face value, quantum coherence is maintained for about a hundred times longer than back-of-the-envelope calculations predicted. Although thermal noise was undoubtedly a factor, more recent calculations[6] suggest that a little bit of noise can actually be good – that is, it can, paradoxically, *boost* the efficiency of energy transfer in the right circumstances (doubling it, in this case). And the photosynthetic system seems to have evolved precisely those 'right circumstances'.

Photosynthesis in plants is more complicated than it is in bacteria, and it is not yet clear whether the quantum effects discovered for the latter are more or less important than in the former. But the Engel–Fleming experiments suggest that quantum-assisted energy transport plays a role in at least one of the basic light-harvesting processes in biology.

SPOOKY BIRDS

'Doth the hawk fly by thy wisdom, and stretch her wings toward the south?'

– Job 39:26

Although bird navigation had been studied informally for many centuries, it wasn't until the early 1700s that ornithologists began keeping systematic records. Johannes Leche, professor of medicine at the University of Turku in Finland, noted that the house martin was the first to arrive in those chilly climes – on 6 May, on average – followed by the barn swallow, on 10 May. (I had no idea birds were so punctual.) Direct observation of migratory patterns was later augmented by ringing the birds and, in recent times, tracking by radar and satellites. Today a great deal of information has been gleaned on this extraordinary phenomenon, including some mind-boggling statistics. Arctic terns, for example, can fly more than 80,000 kilo-metres (49,700 miles) per year, migrating from their breeding grounds in the Arctic all the way to the Antarctic, where they spend the north-ern winters. The blackpoll warbler, which weighs a mere 12 grams, completes a nonstop flight out over the Atlantic Ocean from New England to the Caribbean, where it spends the winter. Some pigeons reliably find their way home after flights of hundreds of kilometres.

How do these birds do it?

Scientists have discovered that birds use a variety of methods to find their way around, taking account of the orientation of the sun and stars as well as local visual and olfactory cues. But this can't be the whole story because some birds can navigate successfully at night and in cloudy conditions. Special interest has focused on the Earth's magnetic field, which is independent of the weather. Experiments with homing pigeons in the early 1970s showed that attaching a mag-net to the bird interfered with its ability to orient correctly. But how, exactly, does a bird sense the Earth's magnetic field, given that it is extremely weak?*

A number of physicists claim that it is quantum physics that en-ables the bird to navigate, by allowing it to *see* the field. Evidently, there has to be some sort of compass inside the creature, coupled to its brain so it can perform in-flight corrections. Tracking that com-pass down hasn't been easy, but in the past few years a plausible candidate has emerged, and it depends on quantum mechanics – in fact, on one of its oddest features.

All fundamental particles of matter possess a property called 'spin'. The idea of spinning bodies is of course familiar and simple enough – the Earth itself spins. Imagine an electron as a scaled-down Earth, shrunk to a point in fact, but retaining its spin. Unlike planets, every electron has *exactly* the same amount of spin, as it does electric charge and mass; it is a basic property they have in common. Of course, electrons go round inside atoms too, and in that manner their speed and direction may vary, depending on which atom and which

---

* Some years ago a physics colleague of mine claimed he could sense north even when blindfolded and disoriented. He attributed it to an ability to detect the Earth's magnetic field. As far as I know, there have been no systematic tests of this unusual ability in humans.

energy level (orbit) they occupy. But the fixed spin I am talking about is intrinsic to the electron, and the full designation is, unsurprisingly, 'intrinsic spin'.

What has this got to do with birds? Well, electrons also possess electric charge (they are the archetypal electrically charged particle, which is why they are called electrons). As Michael Faraday discovered in 1831, a moving electric charge creates a magnetic field. Even if an electron isn't moving from place to place, it is still spinning, and this spin creates a magnetic field around it: all electrons are tiny compasses. So, given that electrons are magnetic as well as electric, they will respond to an external magnetic field much as a compass needle does. That is, the electron will feel a force from the external field that will try to twist it so the poles oppose (north–south). There is, however, a complication. Unlike a compass needle, an electron is spinning. When an external force acts on a spinning body, it doesn't just swing round and line up, it gyrates – a process called 'precession'. That is, the spin axis itself rotates about the line of the applied force. Readers familiar with inclined spinning tops (which precess about the vertical due to the Earth's gravity) will know what I mean.

An isolated electron with nothing more to do than feel the force of Earth's magnetism will execute such a gyration about 2,000 times a second in this case. However, most electrons are employed in atoms, going round and round the nucleus, and the internal electric and magnetic fields of the atom itself, arising from the nucleus and other electrons, swamp the Earth's feeble field, which has negligible effect by comparison. But if an electron is displaced from the atom, it's a different story. That can happen if the atom absorbs a photon. The atom's magnetism weakens rapidly with distance from the nucleus, so the Earth's field becomes relatively more important for the behaviour of the electron. The ejected electron will therefore gyrate differently.

The bird's eyes are being assailed by photons all the time – it's what eyes are for. So here is an opportunity for avian electrons to serve as tiny compasses to steer the bird, but only if there is a way for the bird to know what the ejected electrons are doing. Somehow, the light-disrupted electrons have to engage in some chemistry to send a signal to the bird's brain with information about their activities. The bird's retina is packed with organic molecules; researchers have zeroed in on retinal proteins dubbed 'cryptochromes' to do the job I am describing.[7] When a cryptochrome electron is ejected by a photon, it doesn't cut all its links with the molecule it used to call home. This is where Einstein's spooky action-at-a-distance comes in, used here in the service of the bird. The electron, though ejected from its atomic nest, can still be entangled with a second electron left behind in the protein atom, but, because of their different magnetic environments, the two electrons' gyrations get out of kilter with each other. This state of affairs doesn't last for long; the electron and the positively charged molecule (called a free radical) left behind are stand-out targets for chemical action. (The finger of blame for many medical conditions from diabetes to cancer is pointed at free radicals running amok within cells.) According to the theory of the avian compass, these particular free radicals react either with each other (by recombining), or with other molecules in the retina, to form neurotransmitters, which then signal the bird's brain. This neuro-transmission reaction rate will vary according to the specifics of the spooky link and its mismatched gyrations of the two electrons, which is a direct function of the angle between the Earth's magnetic field and the cryptochrome molecules. So in theory, the bird might actually be able to *see* the magnetic field imprinted on its field of vision. How useful!

Is there any evidence to support this spooky-entanglement story? Indeed there is. A research group at the University of Frankfurt has experimented with captive European robins, which migrate from Scandinavia to Africa, and shown that their direction-finding abilities definitely depend on the wavelengths and intensity of the ambient light, as the theory predicts.[8] Their experiments suggest that the birds combine visual and magnetic data when making decisions on which way to go. The Frankfurt group also tried doubling the ambient magnetic field strength. This initially disrupted the bird's directional sense, but the clever little creatures sorted it all out in about an hour and somehow recalibrated their magnetic apparatus to compensate.

The real clincher came with experiments done at UC Irvine by Thorsten Ritz, in which the birds were zapped by radio frequency (MHz) electromagnetic waves. Beaming the waves parallel to the geomagnetic field had no effect, but when they were beamed vertically the birds became confused.[9] Combining the results of many experiments with different frequencies and ambient light conditions shows the presence of a resonance – a familiar phenomenon in which the energy absorbed by a system spikes at a certain frequency, for example, the opera singer who shatters a wine class when striking the right note. A resonance is exactly what one would expect if the quantum explanation is right, because the radio waves are tuned to typical transition frequencies for organic molecules and would likely interfere with the formation of the all-important spooky entanglement.

The era of quantum ornithology has arrived!

QUANTUM DEMONS UP YOUR NOSE

The sense of smell could provide another terrific example of biological quantum demons at work. Even humans, who don't rank very highly on the scale of olfactory prowess, can distinguish very many different odours. A skilled perfumer (called 'a nose' in the trade) can discriminate between hundreds of subtly different fragrances with a discernment comparable to that of a master wine taster.

How does it work? The basic story is this. Inside the nose are legions of molecular receptors – molecules sporting cavities of many different specific shapes. If a molecule in the air has a complementary shape, it will bind to the corresponding receptor, like a lock and key. Once the docking process happens, a signal is sent to the brain: 'Chanel No. 5!' or similar. Of course, I'm simplifying: odour identification usually involves combining signals from several different receptors. Still, it is clear that olfactory receptors behave like classic Maxwell demons – they sort molecules very precisely by their shapes (rather than speeds – same basic idea) and reject the rest, thus filtering and communicating the information to the brain for the benefit of survival (admittedly, probably not in the case of Chanel, but detecting smoke might qualify).

However, the simple lock-and-key model clearly has shortcomings. Molecules of similar size and shape can smell very different. Conversely, very different molecules can smell similar. It is all very enigmatic. Evidence points to a finer level of discrimination – a demon with sharper senses. An old idea – decades old, in fact – is that, in addition to a molecule's size and shape, its vibrational signature might come into the story. Molecules can (and do) wobble around (thermal agitation, remember), and just as musical instruments have distinctive tones produced by the specific admixture of harmonics, so too do the vibrational patterns of molecules. A buffeted airborne molecule will thus arrive at its nasal docking station jittering about, and a receptor designed to 'pick up the vibes' would provide a useful additional level of discrimination. The mechanism was left vague, however, until 1996, when Luca Turin, then at University College London, proposed that quantum mechanics might be at play; specifically, quantum tunnelling of an electron from odorant molecule to receptor.[10] Turin proposed that the tunnelling electron is coupled to the vibrational states of the odorant molecule (that's a routine mechanism in molecular physics), and further, that the electron energy levels in the receptor molecule are tuned to specific vibrational frequencies of the odorant molecule. The electron that tunnels serves to communicate the docking molecule's identity by absorbing a quantum of energy from the vibration (known to physicists as a phonon – a quantum of sound) and delivering it to the receptor. If the electron's energy matches the receptor's energy level structure, tunnelling is facilitated and a metaphorical light goes on in the nose.

Turin's proposal gave a boost to the vibration theory of smell and offered a possible explanation for otherwise puzzling similarities and differences in smells – it's all down to vibrational patterns rather than the shapes of molecules as such. The theory also offered the advantage of being testable. One check is to try to alter the vibrational modes of the odorant molecule while leaving its chemistry (and also its shape) unchanged. That can be done by substituting various atoms for their isotopes. For example, deuterium, whose nucleus consists of a proton and a neutron, is about twice as heavy as normal hydrogen but chemically identical. Switching a hydrogen atom for a deuterium atom will leave the molecular shape the same but it will alter the vibrational frequencies in an obvious way: heavier atoms move more slowly for the same energy, so vibrate at lower frequencies. And experiments did indeed seem to confirm that the act of deuterating molecules changes the smell, but the results remain controversial and ambiguous.[11] More recently, Turin did the experiment with fruit flies and discovered that they can distinguish between an odorant molecule containing hydrogen and the same molecule containing deuterium. The experimenters also trained the insects to avoid the deuterated molecules, and found that the flies also steered clear of an unrelated molecule with vibrational modes matching that of the deuterated odorant. All this bolsters the theory that quantum tunnelling of vibrational information is key to how flies smell, at least.

QUANTUM BIOLOGY: HERE TO STAY?

> For almost a century, quantum mechanics was like a Kabbalistic secret. But today – largely because of quantum computing – the Schrödinger's cat is out of the bag, and all of us are being forced to confront the exponential Beast that lurks in the current picture of the world.
>
> – Scott Aaronson[12]

Niels Bohr once remarked that anyone who isn't shocked by quantum mechanics hasn't understood it. And shocking it is. While quantum mechanics explains matter brilliantly, it shreds reality. The words 'quantum' and 'weird' inevitably go together. Weird like being in two places at once, or being teleported through barriers or visiting parallel worlds – things that would be utterly bizarre if they happened in daily life. But they occur all the time in the micro-world of atoms and molecules. With so much quantum magic on offer, you'd expect life to be on to it. And it is! As I have described in this chapter, in the last few years evidence has grown to suggest that several important biological processes might be exploiting some aspects of quantum weirdness. They offer tantalizing hints that quantum magic could be all over life. If quantum biology amounts to more than a handful of quirky phenomena, it could transform the study of life as profoundly as molecular biology has done over the past half-century.

When Schrödinger delivered his famous Dublin lectures, quantum mechanics was newly triumphant. It had explained many of the properties of non-living matter. Moreover, it seemed to many physicists of the day that quantum mechanics was sufficiently powerful and sufficiently weird to be able to explain living matter too. In other words, it was hoped that quantum mechanics, or possibly some new 'post-quantum mechanics' still to be worked out, might embed a type of 'life principle' hitherto concealed from us by the sheer complexity of living matter. In his lectures Schrödinger did make use of some routine technical results in quantum mechanics to address the question of how biological information can be stored in a stable form, but he didn't attempt to invoke the sort of weird quantum effects I have described in this chapter to explain life's remarkable properties.

In the decades that followed, few biologists paid much attention to quantum mechanics, most being content to appeal to classical ball-and-stick models of chemistry to explain everything in biology. But in the last few years there has been a surge of interest in quantum biology, although some of the more extravagant claims have given the subject a somewhat suspect status. The key question is whether, if there are indeed non-trivial quantum shenanigans going on in living matter, they are just quirky anomalies or the tip of a quantum iceberg that encompasses *all* life's vital processes. The case studies I have described by no means exhaust all the possible quantum biology effects that have been investigated. The fundamental problem, as will be apparent from the tortuous explanations I have given, is that biology is bewilderingly complex. There is plenty of room for subtle quantum effects to lurk within that complexity but, conversely, there is plenty of room for simple quantum theoretical models to mislead.

The problem in making a case for quantum biology is that the two words, 'quantum' and 'biology', describe domains in tension. Quantum effects are most conspicuous in isolated, cold, simple systems, whereas biology is warm and complex with lots of strongly interacting parts. Quantum mechanics is all about coherence. External disturbances are the enemy of coherence. But as I have explained in the earlier chapters, life loves noise! Biology's demons harness thermal energy to create and to move. Living matter is full of commotion; molecules mill around and bang into each other continually, hook up and shake each other, exchange energy, rearrange their shapes. This pandemonium can't be shut out in live organisms, as it can be in the carefully controlled environment of a physics lab. Nevertheless, there is a fertile middle ground in which noise and quantum coherence coexist for long enough for something biologically useful to happen.[13]

Quantum biology is not just of interest in explaining life, it could also teach quantum engineers some very lucrative tricks. The main focus of quantum engineering today is quantum computing. Consider this statistic. A quantum computer with just 270 entangled particles (entangled = spookily linked) could deploy more information-processing power than the entire observable universe harnessed as a conventional (bit-manipulating) computer. That's because a quantum computer's power rises *exponentially* with the number of entangled components, so a mere 270 entangled subatomic particles have 2^{270} states, which is about 10^{81} (compare 10^{80} atomic particles in the universe). If all those states could be manipulated, it would yield godlike computational power. If a tiny collection of particles has the potential to process mind-numbing amounts of information, would we not expect to see such processing manifested somewhere in nature? And the obvious place to look is biology.

Several years ago there were claims that the molecular machinery implementing the genetic code might be a type of quantum computer.[14] Although there is little supporting evidence that DNA executes true quantum computation, it is possible that some form of quantum-enhanced information processing is going on. Maxwell's demon evades the degrading effects of entropy and the second law by turning random thermal activity into stored bits of information. A quantum Maxwell demon could stave off the same degrading thermal effects that destroy quantum coherence and turn random external noise into stored qubits. If life has evolved such demons able to preserve quantum coherence long enough for the genetic machinery to manipulate the stored qubits, then significant information-processing speed-up might occur. Even a slight boost would confer an advantage and be selected by evolution.

Nevertheless, I should finish this chapter on a cautionary note. All of the putative quantum biology effects I have discussed have been hotly debated.[15] Some early claims were overblown, and more experiments are needed before any definitive conclusions can be drawn. The complexity of biological systems often precludes any simple way to untangle wavelike quantum effects from familiar classical vibrational motion, leaving most of the experiments done so far open to alternative interpretations. The jury, it seems, is still out.[16]

What about the speculation that biology implements actual quantum computation? A long while ago, when dwelling on the profundity of quantum computation, I was struck by a curious thought. It is difficult to imagine any non-living, naturally occurring system doing quantum computation, so I was prompted to ask why, if life is *not* availing itself of this opportunity for exponentially enhanced information processing, the possibility even exists. Why do the laws of physics come with informational capabilities beyond anything that Shannon imagined, if nature hasn't made use of it anywhere in the universe? Has this untapped informational potential sat unexploited by nature for 13.8 billion years just for human engineers to cash in on?

I fully realize that what I have just written is in no sense a scientific argument; it is a philosophical (some might say theological) one. I raise it because in my experience as a theoretical physicist I have found that if well-established physical theory predicts that something is possible, then nature invariably seems to make use of it. One need think no further than the Higgs boson, predicted by theory in 1963 and discovered really to exist in 2012. Other examples include anti-particles and the omega minus particle. In all cases there was a well-defined place for such a thing in nature and, sure enough, they are out there. Of course, there are many speculative theories that make predictions *not* borne out by experiment, so my argument is only as good as the reliability of the theory concerned. But quantum mechanics is *the most* reliable theory we have, and its predictions are almost never questioned. Quantum mechanics has a place for exponential godlike information management; has nature overlooked to fill it? I don't think so.

6

Almost a Miracle

'How remarkable is life? The answer is: very. Those of us who deal in networks of chemical reactions know of nothing like it.'

– George Whitesides[1]

The universe abounds in complexity, from everyday systems such as turbulent streams and snowflakes to grand cosmic structures like nebulae and spiral galaxies. However, one class of complex systems – life – stands out as especially remarkable. In his Dublin lectures Schrödinger identified life's ability to buck the trend of the second law of thermodynamics as a defining quality. Living organisms achieve this entropy-defying feat by garnering and processing information and directing it into purposeful activity. By coupling patterns of information to patterns of chemical reactions, using demons to achieve a very high degree of thermodynamic efficiency, life conjures coherence and organization from molecular chaos. One of the greatest outstanding questions of science is how this unique arrangement came about in the first place.

How did life begin? Because living matter has both a hardware and a software aspect – chemistry and information – the problem of origins is doubly difficult. In a curious historical coincidence, just three weeks after Crick and Watson's famous paper on the double-helix structure of DNA appeared in *Nature*, the 15 May 1953 edition of *Science* carried an article by a little-known chemist named Stanley Miller. Entitled 'A production of amino acids under possible primitive Earth conditions', it was subsequently hailed as a trailblazer for attempts to re-create life in the laboratory.[2] Miller put a mixture of common gases and some water in a flask and sparked electricity through it for a week, producing a brown sludge. Chemical analysis showed that this simple procedure had succeeded in making some of the amino acids life uses. It looked as if Miller had taken the first step on the long road to life with little more than a bottle of gas and a pair of electrodes. The conjunction of these two papers – one on life's giant informational molecule, the other on its simple chemical building blocks – aptly symbolizes the central problem of biology: what came first, complex organic chemistry or complex information patterns? Or did they somehow bootstrap each other into existence in lockstep? What is clear is that chemistry alone falls short of explaining life. We must also account for the origin of organized information patterns. And not just information: we also need to know how *logical operations* emerged from molecules, including digital information storage and mathematically coded instructions, implying as they do *semantic* content. Semantic information is a higher-level concept that is simply meaningless at the level of molecules. Chemistry alone, however complex, can never produce the genetic code or contextual instructions. Asking chemistry to explain coded information is like expecting computer hardware to write its own software. What is needed to fully explain life's origin is a better understanding of the organizational principles of information flow and storage and the manner in which it couples to chemical networks, defined broadly enough to encompass both the living and non-living realms. And the overriding question is this: can such principles be derived from known physics or do they require something fundamentally new?

IN THE BEGINNING . . .

Francis Crick once described the origin of life as 'almost a miracle, so many are the conditions which would have had to have been satisfied to get it going'.[3] And it's true that the more 'miraculous' life appears to be, the harder it is to figure out how it can have started. In 1859 Charles Darwin's magnum opus *On the Origin of Species* first appeared. In it he presented a marvellous account of how life has evolved over billions of years from simple microbes to the richness and complexity of Earth's biosphere today. But he pointedly left out of his account the question of how life got started in the first place. 'One might as well speculate about the origin of matter,' he quipped in a letter to a friend.[4] Today, we are not much further forward. (Except we do more or less understand the origin of matter in the Big Bang.) My earlier chapters will have convinced the reader, I hope, that life is not just any old phenomenon but something truly special and peculiar. How, then, can we account for the transition from non-life to life?

The enigma of life's origin is actually three problems rolled into one: when, where and how did life begin? Let me first deal with when. The fossil record can be traced back about 3.5 billion years, to a geological epoch known as the Archaean. It's hard to find many rocks this old, let alone spot any fossils therein. One outcrop of Archaean chert has, however, been intensively studied. It is located in the Pilbara region of Western Australia, about a four-hour drive into the bush from the town of Port Headland. The terrain is rugged, sparsely vegetated and scoured by mostly dry riverbeds prone to flash flooding. The hills here are a rich red hue and rocky outcrops harbour important traces of ancient microbial activity. Dating these rocks indicates that our planet was already hosting a primitive form of life within a billion years of its formation. The most persuasive evidence comes from curious geological features known as stromatolites. They appear as ranks of wavy lines or little humps decorating the exposed rock surfaces. If the interpretation is correct, these features are remains of what, 3.5 billion years ago, were microbe-covered mounds, created by successive microbial colonies depositing mats of grainy material on the exposed surfaces, layer by layer. There's just a handful of places on Earth where one may today see similar stromatolite structures complete with their living microbial residents. Most geologists are confident that the Pilbara stromatolites (and others in younger rocks around the world) are the fossil relics of something similar, dating from the far past. The same Pilbara geological formation contains additional signs of life in remnants of an ancient reef system and a few putative individual fossilized microbes. It's hard to tell from the shapes alone that the 'fossils' are more than merely marks in a rock;

any organic material has long gone. However, very recently the bio-genic interpretation received a boost.[5] About 99 per cent of the carbon on Earth is in the form of the lighter isotope C^{12}. Life favours this lighter form so fossils usually possess a slight additional abundance of it. An analysis of the Pilbara rocks shows that the carbon isotope ratio is correlated with the physical shapes of the marks, as it would be if these were fossils of different microbial species. The results are hard to explain non-biologically.

The evidence of the Pilbara tells us that life was established on Earth by 3.5 billion years ago, but it gives little clue as to when life may have actually started. It's possible that all older traces of biologi-cal activity have been obliterated by normal geological processes, and by the bombardment of our planet by large asteroids that occurred until about 3.8 billion years ago – the same bombardment that cra-tered the moon so thoroughly. The problem is a lack of older rocks. Greenland has some dating back more than 3.8 billion years, with hints of biological modification, but they aren't decisive. Neverthe-less, Earth itself is only 4.5 billion years old so life has been present here for at least 80 per cent of its history.

WHERE DID LIFE BEGIN?

Although the when part of the origin question can at least be bounded, it is much harder to guess *where* life first appeared. I don't mean the latitude and longitude as such but the geological and chemical setting.

The first thing to say is that there is no compelling evidence that terrestrial life *started* on Earth. It may have got going elsewhere and come to Earth ready-made. For example, it may have begun on Mars, which before about 3.5 billion years ago was warmer and wetter than today, and more Earth-like. In some respects, Mars offered a *more* favourable environment for pre-biotic chemistry. For example, the effects of the asteroid bombardment may have been less severe and the chemical make-up of the red planet was better for driving metab-olism. Obviously, there would have to be a way for life to spread from Mars to Earth, and there is. The bombardment by asteroids and comets, which was severe in the early history of the solar system (but has never entirely ceased), is capable of blasting vast amounts of rock into space, much of which goes into solar orbit. A fraction of ejected Mars rocks will eventually fall to Earth (and vice versa – terrestrial rocks go to Mars). Rocks from Mars, which fall as meteorites, have been collected from all over the world; my university has several. Over the history of our planet, trillions of tons of Martian material have come here. Ensconced in a chunk of rock, a microbe could withstand the harsh conditions of outer space. The greatest hazard in crossing the interplanetary void is radiation, but even a moderately sized rock would screen most of that out. It has been estimated that some hardy radiation-resilient microbes could survive for millions of years inside space rocks, easily long enough to reach Earth and seed it with Martian life. The same scenario works in reverse: viable terrestrial microbes can reach Mars. What this means is that Earth and Mars are not quarantined from each other. Cross-contamination by microbial life could have been going on throughout history. This makes it hard to be sure that life on Earth began here and not there. It is possible, but less likely, that life reached Earth from Venus, which is now very hostile to life but may have been more congenial billions of years ago. Another possibility, taken seriously in some quarters, is that life was originally incubated in a comet and delivered to Earth either by a direct impact or, more probably, from cometary dust that filtered down after a near-miss.

Shifting the cradle of life from Earth to somewhere else doesn't much advance the more important question of what geological setting would be conducive to producing life. Many scenarios have been touted: deep ocean volcanic vents, drying lagoons, pores in sub-ocean rocks ... the list is long. About the only thing everyone agrees on is that oxygen gas would have been a frustrating factor. Today, complex organisms require oxygen for their metabolism, but this was a late development. On Earth, there was very little free oxygen in the atmosphere before about 2 billion years ago, and present levels were not attained until within the last billion years. Oxygen may feel good to breathe, but it is a highly reactive substance that attacks and breaks down organic molecules. Aerobic life has evolved all sorts of mechanisms to cope with it (such as anti-oxidants). Even so, reactive oxygenic molecules regularly damage DNA and cause cancer. When it comes to the origin of life, free oxygen is a menace.

The elements essential to life do include oxygen, of course, but also hydrogen, nitrogen, carbon, phosphorus and sulphur. The truly essential element is carbon, the basis of all organic chemistry, and an ideal choice because of the limitless variety of complex molecules it can form. Chemists envisage the first steps towards life to have taken place where there was a good supply of carbon (for example, from carbon dioxide) and also hydrogen, perhaps free, or as a constituent of methane or hydrogen sulphide. A popular suggestion of locale is in the vicinity of volcanic vents under the ocean, where sulphur is also in good supply and the rocky surfaces offer all sorts of possible catalysts. Scientists have focused on such places because of the discovery of rich ecosystems clustering near deep subsea vents, perilously close to the scalding high-pressure effluent spewing from the volcanic depths. The primary producers at the base of the food chain are heat-loving microbes known as 'hyperthermophiles'; some of these dare-devil organisms have been found thriving in water above 120°C. (The water doesn't boil at these temperatures because of the intense pressure.) Nobody expected to find life in the dark depths, and certainly not in the pressure-cooker conditions near volcanoes. But the surprise didn't end there. One of the most astonishing discoveries in biology in recent decades is that life is not restricted to the Earth's surface or the oceans but extends deep underground, both on land and beneath the sea bed. The full extent of this subterranean biosphere is still being mapped, but microbes have been found living several kilometres down, inside rock (the South African extremophiles I mentioned in Chapter 2 are one such example).*

There has been a lively debate about whether life *started* deep inside the Earth's crust or whether it infiltrated the subsurface after first establishing itself above (or having arrived from Mars perhaps). Genetic sequencing has shown that hyperthermophiles occupy the deepest and, by implication, oldest branches on the tree of life, suggesting that heat resilience is a very ancient feature of terrestrial

* When the astrophysicist Thomas Gold suggested in the late 1980s that there may exist a deep, hot biosphere, he received nothing but ridicule. Yet he was absolutely right.

biology, but that does not necessarily mean the first living things were hyperthermophiles. Life may have started somewhere cooler and then diversified, with some microbes evolving the necessary heat-damage-repair mechanisms to enable them to colonize the hot subsurface or the sea bed near ocean vents. Because the early bombardment probably included impacts by objects big enough to heat-sterilize large areas of the surface (if not the whole planet), then only the heat-loving subterranean microbes would have survived. They would thus represent a genetic bottleneck rather than representatives of the very first life forms. At this stage, it's impossible to know.

Armed with a basic notion of the chemical setting (no oxygen!), scientists have spent decades trying to re-create conditions in the laboratory that might illuminate the first chemical steps on the long pathway to life, following Miller's pioneering efforts in 1953. Many subsequent pre-biotic synthesis experiments have been done but, to be honest, they don't get very far, in spite of the dedication and ingenuity of the scientists. By the standards of biological molecular complexity, these attempts barely make it to first base.

There is a more fundamental reason why efforts to cook up life in the lab are unlikely to solve the mystery of life's origin. As I have stressed in this book, the distinctive character of life is its ability to store and process information in an organized manner. Of course, life also requires complex chemistry; organic molecules form the substrate in which life performs its software feats. But it's only half the story – the hardware half. Obviously, there *was* a chemical pathway from non-life to life, even if we have scant idea what it was, but the actual chemical steps may not have been as important as the really critical transition: the one from inchoate molecular mayhem to organized information management. How did *that* happen?

HOW DID LIFE BEGIN?

I've left the hardest problem – how life began – to last. The short answer is, nobody knows how life began! It's worse: nobody even knows how to go about estimating the odds for it to happen. But a lot hinges on the answer. If life starts easily, the universe should be teeming with it. Furthermore, if terrestrial life is the product of a universe that embeds some form of life principle in its basic laws, then the place of human beings in the great cosmic scheme would be profoundly different than if we were the products of a freak chemical accident.

As I have mentioned, a basic unknown about the pathway from non-life to life is whether it was a long, steady slog up a pre-biotic version of Mount Improbable, or whether it took place in fits and starts, with long periods of stasis interrupted by great leaps forward (or upward, in this metaphor). Given that Mount Improbable is so incredibly high, it won't do for a chemical mixture to attain a toehold in the foothills only to slide back down again. There has to be some sort of ratcheting effect to lock in the gains and limit the losses while the system hangs out for the next step. But ideas like this, which seem sensible enough, run into the problem of teleology. A chemical soup doesn't know it's trying to make life – a chemical soup doesn't know anything at all – so it won't act to protect its hard-won complexity from the ravages of the second law of thermodynamics. Scenarios in which chemistry 'strives' towards life are patently absurd. The same problem doesn't occur once life gets going, because natural selection can ratchet up the gains and DNA storage can lock them in. But chemistry without natural selection has no recourse to such mechanisms.*

The backsliding problem afflicts almost all studies of the complexification pathway to life. There are many clever experiments and theoretical analyses demonstrating the spontaneous formation of complexity in a chemical mixture, but they all hit the same issue: what happens next? How does a chemical broth build on some spontaneously emerging complexity to then ramp up to something even more complex? And on and on, until the summit of the pre-biotic Mount Improbable is reached? The most promising break-out from this straitjacket comes from work on 'autocatalytic' chemical cycles.

---

* There is a long history of what might be called 'molecular Darwinism' in which 'naked' molecules are able to replicate with varying efficiency and natural selection filters out the best. The so-called RNA world theory falls into this category. Though these studies are instructive, they are very contrived and require carefully managed human intervention (e.g. to prepare materials, to do the selecting) to accomplish anything. Relevance to the natural world is far from obvious.

The idea here is that certain molecules, say A and B, react to make other molecules, C, that happen to serve as catalysts to accelerate the production of A and B. There is thus a feedback loop: groups of molecules catalyse their own production. Scaling this up, there could be a vast network of organic molecules forming a quasi-stable system of autocatalysis, with many interlocking feedback loops, combining in a tangled web of reactions that is self-sustaining and robust.[6] All this is easy to state in words, but are there such chemical systems out there? Yes, there are. They are called living organisms and they deploy all the aforementioned features. But now we are going round in circles, because we want to ascertain how all this marvellous chemistry can take place *before* life. We can't put the solution by hand into the problem we are trying to solve and then claim to have solved it.

And the problem is more severe than I have stated. One of the informational hallmarks of life is the way it manages digital information using a mathematical code. Recall that triplets of letters (A, G, C, T) stand for specific amino acids from among the toolkit of twenty or so used to make proteins. The coded instructions transported from DNA to the protein assembly machinery (ribosomes, tRNA, and so on) are a prime example of Shannon's information theory at work, with the instructions playing the role of the message, the communication channel being the watery innards of the cell and the noise being thermal or chemical mutational damage to the mRNA en route.

An explanation for the origin of life as we know it has to include an explanation for the origin of such digital information management and – especially – the origin of the code. (It doesn't have to be the actual code known life uses, but the origin of *some* sort of code needs an explanation.) This is a tough, tough problem. Biochemists Eugene Koonin and Artem Novozhilov call it 'the most formidable problem of all evolutionary biology', a problem that 'will remain vacuous if not combined with understanding of the origin of the coding principle itself and the translation system that embodies it'. They don't think it will be solved any time soon:

> Summarizing the state of the art in the study of the code evolution, we cannot escape considerable skepticism. It seems that the two-pronged fundamental question: 'why is the genetic code the way it is and how

did it come to be?', that was asked over 50 years ago, at the dawn of molecular biology, might remain pertinent even in another 50 years. Our consolation is that we cannot think of a more fundamental problem in biology.[7]

It's certainly correct that biologists have puzzled over the origin of the code for a long time. A popular proposed solution is that primitive life didn't use a code, that what we have today represents a sort of software upgrade which evolved later once natural selection kicked in. The so-called RNA world theory has developed along these lines. Since it was discovered in 1982 that RNA can both store information *and* catalyse RNA chemical reactions (not as well as proteins, but maybe well enough to pass muster) biochemists have wondered whether an RNA soup could 'discover' replication with variation and selection all on its own, with proteins coming later. Even if this explanation is along the right lines, however, it is all but impossible to estimate the odds of such a scenario being played out on a planet. It's easy to imagine those odds being exceedingly adverse.

Fifty years ago the prevailing view among biologists was that the origin of life was a chemical fluke, involving a sequence of events that collectively was so low in probability that it would be unlikely to recur anywhere else in the observable universe. I have already quoted Crick. His French contemporary Jacques Monod criticized the idea that life is somehow 'waiting in the wings' ready to burst forth whenever conditions permit. He summarized the prevailing view among scientists as follows: 'the universe is not pregnant with life', and therefore 'Man knows at last that he is alone in the indifferent immensity of the universe, whence which he has emerged by chance.'[8] George Simpson, one of the great neo-Darwinists of the postwar years, dismissed SETI, the search for intelligent life beyond Earth, as 'a gamble at the most adverse odds with history'.[9] Biologists such as Monod and Simpson based their pessimistic conclusions on the fact that the machinery of life is so stupendously complex in so many specific ways that it is inconceivable it would emerge more than once as a result of chance chemical reactions. In the 1960s to profess belief in extraterrestrial life of any sort, let alone intelligent life, was tantamount to scientific suicide. One might as well have expressed a belief in fairies. Yet by the 1990s sentiment had swung. The Nobel prize-winning biologist Christian de Duve, for example, described the universe as 'a hotbed of life'. He was so convinced that life would emerge wherever it had a chance he called it 'a cosmic imperative'.[10] And that seems to be the fashionable view today, where appeal is often made to the huge number of habitable planets deemed to be out there. Consider, for example, the sentiments expressed by Mary Voytek, former head of NASA's Astrobiology Institute: 'With all the other planets around all the other stars, it's impossible to imagine that life would not have arisen somewhere else.'[11] Well, it's not only possible, it's actually rather easy to imagine. Suppose, for example, the transition from non-life to life involved a sequence of a hundred chemical reactions, each requiring a particular temperature range (for example, 5–10°C for the first, 20–30°C for the second, and so on). Perhaps the transition also demanded tightly constrained pressure, salinity and acidity ranges, not to mention the presence of a host of catalysts. There might be only one planet in the observable universe where the necessary dream run of conditions occurred. My conclusion: *habitability* does not imply *inhabited*.

Why is it now scientifically respectable to search for life beyond Earth, whereas it was taboo even to talk about it half a century ago? There is no doubt that the discovery of so many extra-solar planets has provided astrobiology with a huge fillip. However, though no planets outside the solar system had been detected in the sixties, most astronomers nevertheless supposed they were there. A further point astrobiologists now make is the discovery of organic molecules in space, providing evidence that abundant 'raw material' for life is scattered throughout the universe. That may be so, but there is a vast complexity gulf separating simple building blocks such as amino acids from a metabolizing, replicating cell. The fact that the first small step across that chasm might have already been taken in space is almost irrelevant. Yet another reason given for the current optimism about life beyond Earth is the recognition that some types of organisms can survive in a much wider range of physical conditions than was recognized in the past, opening up the prospect for life on Mars, for example, and generally extending the definition of what constitutes an 'Earth-like' planet.

But this at most amounts to a factor of two or three in favour of the odds for life. Set against that is the *exponentially* small probability that any given complex molecule will form by random assembly from a soup of building blocks. In my opinion, we remain almost completely in the dark about how life began, so attempts to estimate the odds of it happening are futile. You cannot determine the probability of an unknown process! We cannot put any level of confidence – *none at all* – on whether a search for life beyond Earth will prove successful.

There is one argument for the ubiquity of life that does carry some force. Carl Sagan once wrote, 'The origin of life must be a highly probable affair; as soon as conditions permit, up it pops!'[12] It is true that life was here on Earth very soon (in geological terms) after our planet became congenial for it. Therefore, reasoned Sagan, it must start readily. Unfortunately, the conclusion doesn't necessarily follow. Why? Well, had life *not* started quickly, there wouldn't have been time for it to evolve as far as intelligence before Earth became uninhabitable, fried to a crisp by the steadily increasing heat of the sun. (In about 800 million years the sun will be so hot it will boil the oceans.) Put simply, unless life was quick off the mark, we wouldn't be here today discussing it. So, given that our own existence on this planet *depends* on life forming here, it's entirely possible that the origin of terrestrial life was an extreme outlier, an immense fluke.

MAKING LIFE IN THE LAB

Sometimes it is suggested that, if we could only make life in the laboratory, it would demonstrate clearly that it isn't a fluke but can start up easily. Media reports often give the misleading impression that life has *already* been created in the lab, often with the moral subtext that 'playing God' in this manner might invite Frankenstein-like comeuppance. For example, on 20 May 2010 Britain's *Daily Telegraph* featured a headline 'Scientist Craig Venter creates life for first time in laboratory sparking debate about "playing God"'. This is deeply misleading. The misunderstanding comes down to the ambiguous term 'create'. In one sense, humans have been creating life for centuries, the most obvious example being dogs. Dogs are artificial animals produced from wolves by generations of cross-breeding and careful selection. Twenty thousand years ago there were wolves but no Great Danes or chihuahuas. In more recent years genetic engineering techniques such as gene transplantation have enabled many novel organisms to be created, including a variety of GM foods. New technology known as CRISPR enables genomes to be rewritten more or less to order. What Venter and his colleagues did was brilliant and deservedly attention-grabbing. He took a simple bacterium (*mycoplasma genitalium*) and replaced its DNA with a customized version. In other words, Venter kept almost all the hardware (the cell) and just switched the software (the DNA). The mycoplasma obligingly booted up the new software and ran the re-engineered genetic instructions; the new organism was dubbed *mycoplasma laboratorium*. The computer equivalent would be like buying a PC and reinstalling your own version of the operating system with a few designer embellishments added. Would that amount to creating a computer? Not really. Loose talk of creating life in the lab conflates chemistry with information, hardware with software. The main point is that re-engineering existing life, which is what Venter did, is very far indeed from making life from scratch.

Occasionally, there are media reports suggesting that even that more ambitious goal is close at hand. On 27 July 2011 *The New York Times* reported, beneath the dramatic headline 'It's alive! It's alive! Maybe right here on Earth', that 'a handful of chemists and biologists . . . are using the tools of modern genetics to try to generate the Frankensteinian spark that will jump the gap separating the inanimate and the animate. The day is coming, they say, when chemicals in a test tube will come to life.' The reporting is accurate enough. However, the definition of life being employed in the said experiments is extremely loose: a mixture of molecules that can make copies of themselves with occasional errors (mutations). In terms of chemistry, this work is without doubt an outstanding accomplishment and provides a helpful piece in the jigsaw puzzle of life. But, as the experimenters would be the first to admit, their molecular replication system is a far cry from a living cell with an autonomous existence.

The fundamental problem is not the simplicity of the components in these experiments. It is something far deeper. To attain even the modest successes announced so far requires special equipment and technicians, purified and refined substances, high-fidelity control over physical conditions – and a big budget. But above all, it needs an intelligent designer (aka a clever scientist). The organic chemist must have a preconceived notion of the entity to be manufactured. I'm not denigrating the scientists involved or the glittering promise of the field of synthetic biology, only its relevance to the *natural* origin of life. Astrobiologists want to know how life began *without* fancy equipment, purification procedures, environment-stabilizing systems and – most of all – without an intelligent designer. It may turn out that life is indeed easy to make in the lab but would still be exceedingly unlikely to happen spontaneously in the grubby and uncertain conditions available to Mother Nature. After all, organic chemists can readily make plastics, but we don't find them occurring naturally. Even something as simple as a bow and arrow is straightforward for a child to make but would never be created by an inanimate process. So just because *we* might (one day) find life easy to create does not of itself demonstrate a cosmic imperative.

What would swing the debate is if, by synthesizing life many times and in many different ways, scientists uncovered certain common principles which could then be applied to real-world conditions. And that would open up the profound question of whether such principles already lurk within the corpus of scientific knowledge or require something entirely new. Schrödinger was open-minded on this matter: 'We must therefore not be discouraged by the difficulty of interpreting life by the ordinary laws of physics. For that is just what is to be expected from the knowledge we have gained of the structure of living matter. We must also be prepared to find a new type of physical law prevailing in it,' he wrote.[13] I agree with Schrödinger. I believe there are new laws and principles that emerge in information-processing systems of sufficiently great complexity, and that a full explanation for life's origin will come from a detailed study of such systems. I shall return to this speculative theme in the Epilogue.

Meanwhile, all is not hopeless on the observational front.

Box 12: Is life a *planetary* phenomenon?

Life as we know it has three fundamental features: genes, metabolism and cells. Clearly, they didn't all spring into existence at once, and one of the challenges in origin-of-life research is to decide what came first. Among the three, cells are the easiest to form. There are many substances that spontaneously produce cellular structures, so an early speculation is that legions of small vesicles were available on the early Earth to serve as natural 'test tubes' in which nature might experiment with complex organic chemistry. Cells also fulfil another critical function. Darwinian evolution needs a unit to select on, and cells fit the bill. Even a non-living blob can reproduce after a fashion by fissioning into two smaller blobs, opening the way for a population of similar entities to serve an evolutionary role. Without the existence of *individuals* the original version of Darwinism is meaningless.

Recently, an opposing view has gained attention. Perhaps cells came later, after complex chemistry had already established something like metabolic cycles and networks. This chemical self-organization could occur in 'the bulk' – in the open oceans, say – on a large scale. Once the metabolic processes became self-sustaining and self-reinforcing, the way would have been open for fragmentation into individual units, culminating in what we would today recognize as living cells. It would be a top-down approach to the origin of life. The pre-cellular phase may have been restricted to thermodynamically favourable environments, such as deep ocean volcanic vents, or it may have encompassed the entire planet. Eric Smith and the late Harold Morowitz paint a picture of life as an essentially geological or planetary phenomenon, in which the geochemistry of the early Earth co-evolves with pre-life. Eventually, what we call life emerges, they conjecture, from a sort of planetary phase transition.[14] It is an intriguing hypothesis.

A SHADOW BIOSPHERE

Suppose chance played only a subordinate role in incubating life and that the process was more 'law-like', more of an imperative, as de Duve expressed it. Is it possible that the blueprint for life is somehow embedded in the laws of physics and is thus an *expected* product of an intrinsically bio-friendly universe? Perhaps. The trouble is, these musings are philosophical, not scientific. What sort of law would imply that life arises more or less automatically wherever conditions permit? There is nothing in the laws of physics that singles out 'life' as a favoured state or destination. All the laws of physics and chemistry discovered so far are 'life blind' – they are universal laws that care nothing for biological states of matter, as opposed to non-biological states. If there is a 'life principle' at work in nature, then it has yet to be discovered.

For the sake of argument, let me join the ranks of the optimists who say that life starts easily and is widespread in the cosmos. If life is inevitable and common, how might we obtain evidence for it? If we found a second sample of life (on another planet, a moon, a comet) that we could be sure had arisen from scratch independently of known life, the case for de Duve's cosmic imperative would be instantly and dramatically confirmed. In my view, the most promising place to search for a second genesis is right here on our own planet. If life does indeed get going easily, as so many scientists fervently believe, then surely it should have started many times on Earth. Well, how do we know it didn't? Has anybody actually looked?

Consider this scenario: life emerges on planet Earth 4 billion years ago. Ten million years later a huge asteroid strikes, releasing so much heat that the oceans boil and the surface of the planet is sterilized. The massive blow would not, however, destroy all life. Vast quantities of rock would be spewed into space, some of it containing Earth's first tiny inhabitants. The microbial cargo could survive for many millions of years, orbiting the sun. Eventually, some of this material would find its way back to Earth and fall as meteorites, bringing life home. But meanwhile, in the few million years since the cataclysmic impact, life has got going a second time (it starts easily, remember), so when the ejected material returns there are now *two* forms of life on our planet. Because the barrage of huge objects continued for 200 million years, this same scenario could have played out many times, so that when the bombardment finally abated there may have been dozens of independently formed organisms cohabiting our planet. The fascinating question is, might at least one of these examples of life-as-we-don't-know-it have survived to the present day? Almost all life on Earth is microbial, and you can't tell by looking what makes a microbe tick. You have to delve into its molecular innards. So might there be, intermingled with the microbes representing 'our' form of life, representatives of this 'other' life – it would be truly alien life, in the sense of being descended from an independent genesis. The existence of an alien microbial population has been dubbed a 'shadow biosphere', and it carries the intriguing possibility that there might be alien life right under our noses – or even *in* our noses – overlooked so far by microbiologists.[15]

Identifying shadow life would be a challenge. My colleagues and I have come up with some broad strategies, which I explained in *The Eerie Silence*. For example, we could search in places where conditions are so extreme they lie beyond the reach of all known life – even of the extremophile kind – such as near volcanic vents beneath the sea in regions of the effluent where the temperature exceeds 130°C. On the other hand, if shadow life is intermingled with known life, the task of identifying it would be harder. A chemical agent that killed or slowed the metabolism of all known life might enable a minority population of shadow-life microbes to flourish and so stand out. A few scientists have made a start along these lines but, considering the momentous consequences of such a discovery, it is surprising how little attention it has attracted. All it would take to settle this question is the discovery of a single microbe – just one – which represents life, but not as we know it. If we had in our hands (or rather under our microscopes) an organism whose biochemistry was sufficiently unlike our own that an independent genesis was unavoidable, the case for a fecund universe would be made. If life can happen twice, it can surely happen a zillion times. And that single alien microbe doesn't have to be on some far-flung planet; it could be here on Earth. It could be discovered tomorrow, upending our vision of the cosmos and mankind's place within it and greatly boosting the prospect that intelligent life may be out there somewhere.

Looking back over the past 3.5 billion years, the origin of life was the first, and most momentous, transformation. However, the history of evolution contains other major transitions, critical steps without which further advance would be impossible.[16] It took a billion years or so after life began for the next major transition: the arrival of eukaryotes. Another big step was sex. Later came the leap from unicellularity to multicellularity. What prompted these further transformations to occur? Are there any common underlying features? Eukaryogenesis, sex and multicellularity: all involved marked physical alterations. But the true significance lay not with changes in form or complexity but with the concomitant reorganization of *informational architecture*. Each step represented a mammoth 'software upgrade'. And the biggest upgrade of all began about 500 million years ago with the appearance of a primitive central nervous system. Fast-forward to today, and the human brain is the most complex information-processing system known. From that system stems what is undoubtedly the most astonishing phenomenon of all in life's magic puzzle box – consciousness.

7

The Ghost in the Machine

*'Regarding the nature of the very foundation of mind, con-
sciousness, we know as much as the Romans did: nothing.'*
— *Werner Loewenstein*[1]

Thirteen years after delivering his Dublin lectures, Erwin Schrödinger
returned to the subject of life in a series of presentations at Cambridge
University entitled 'The physical basis of consciousness'.[2] Focusing on
the question 'What kind of material process is directly associated
with consciousness?', he proceeded to give a physicist's eye-view of
this most extraordinary of phenomena. Among life's many baffling
properties, the phenomenon of consciousness leaps out as especially
striking. Its origin is arguably the hardest problem facing science
today and the only one that remains almost impenetrable even after
two and a half millennia of deliberation. If Schrödinger's question
'What is life?' has proved hard enough to answer, 'What is mind?' is
an even tougher nut to crack.

An explanation of mind, or consciousness, is more than an academic challenge. Many ethical and legal questions hinge on whether,
or how much, consciousness is present in an organism. For example,
opinions about abortion, euthanasia, brain death, vegetative states
and locked-in syndrome may depend on the extent to which the
subject is conscious.* Is it right to artificially prolong the life of a

---

* A high-profile case occurring at the time of writing was that of Charlie Gard, a
baby born with an apparently incurable syndrome that left him largely unrespon-
sive. A legal decision was made to terminate life support rather than permit
experimental treatment.

permanently unconscious human being? How can we tell if an unre-
sponsive stroke victim might actually be aware of their surroundings
and in need of care? Animal rights involving definitions of cruelty are
often based on very informal arguments about whether and when an
animal can suffer or 'feel pain'.* Added to these concerns there is the
emerging field of non-biological intelligence. Can a robot be conscious, and if so does it have rights and responsibilities? If we had an
accepted definition of 'degree of consciousness' based on a sound scientific theory, then perhaps we could make better judgements about
such contentious matters.

What is lacking is a comprehensive theory of consciousness. In
Western societies there is a popular notion that the conscious mind
is an entity in its own right. It is a view often traced back to the
seventeenth-century French philosopher-scientist René Descartes,
who envisaged human beings as made of two sorts of things: bodies
and minds. He referred to *res extensa* (roughly speaking, material
stuff) and *res cogitans* (wispy mind-stuff). In popular Christian culture the latter concept has sometimes become conflated with the soul,
an immaterial extra ingredient that believers think inhabits our bodies and drifts off somewhere when we die. Modern philosophers (and
theologians, for that matter) generally take a dim view of 'Cartesian
dualism' as Descartes' 'separation of powers' is known, preferring
to think of human beings as unitary entities. In 1949 the Oxford
philosopher Gilbert Ryle coined the pejorative phrase 'the ghost in
the machine' to describe Descartes' position (which he called 'the
official view' of mind). He derisorily drew an analogy between our
immaterial minds controlling our mechanical bodies with, say, a car
under the control of a driver.[3] Ryle argued that this mystical 'dogma'
was not only wrong in fact but deeply flawed conceptually. Yet in
the popular imagination, the mind is still regarded as some sort of

---

* There was a famous legal test in Britain in 1973: the case of the Eyemouth prawns.
A sixteen-year-old girl was charged with cruelty to animals for cooking live prawns
on a hotplate. The case was eventually dropped, but not before it attracted the attention of the Soviet media. At that time Britain was going through a period of industrial
turmoil and economic decline. Moscow cited the prawn case as an example of Western decadence: how could the British people be preoccupied with such trivia when
the workers were in revolt against the collapsing capitalist system?

nebulous ghost in the machine. In this book I have argued that the concept of information can explain the astonishing properties of living matter. The supreme manifestation of biological information processing is the brain, so it is tempting to suppose that some aspect of information will form a bridge between mind and matter, as it does between life and non-life. Swirling patterns of information do not constitute a 'ghost' any more than they constitute a 'life force'. Yet the manipulation of information by demon-like molecular structures is perhaps a faint echo of the dualism that Ryle derided. It is, however, a dualism rooted, not in mysticism, but in rigorous physics and computational theory.

IS ANYONE AT HOME?

To get started, let's consider what we mean when we talk about consciousness in daily life. Most of us have a rough and ready definition: consciousness is an awareness of our surroundings and our own existence. Some people might throw in a sense of free will. We possess mental states consisting of feelings, thoughts and sensations, and somehow our mental world couples to the physical world through our brains. And that's about as far as it goes. Attempts to define consciousness more precisely run into the same problems as attempts to define life but are far more vexing. The mathematician Alan Turing, famous for his work on the foundations of computing, addressed this question in a paper published in 1950 in *Mind*.[4] Asking the loaded question 'Can machines think?', Turing pre-figured much of today's hand-wringing over the nature of artificial intelligence. His main contribution was to define consciousness by what he called 'the imitation game',* often referred to as 'the Turing test'. The basic idea is that if someone interrogates a machine and cannot tell from the answers whether the responses are coming from a computer or another human being, then the computer can be defined as conscious.

Some people object that just because a computer may convincingly simulate the appearance of consciousness doesn't mean it *is* conscious;

---

* That being the title of the recent movie about Turing's life.

the Turing test attributes consciousness purely by analogy. But isn't that precisely what we do all the time in relation to other human beings? Descartes famously wrote, 'I think, therefore I am.' But although I know my own thoughts, I cannot know yours without being you. I might infer from your behaviour, by analogy with mine, that 'there's somebody at home' inside your body, but I can never be sure. And vice versa. The best I can say is 'you look like you are thinking so you look like you exist'. There is a philosophical position known as solipsism that denies the existence of other minds. I won't pursue it here because if you, the reader, don't exist, then you won't be interested in my arguments for solipsism and I will be wasting my time.

Philosophers have spent centuries trying to link the worlds of mind and matter, a conundrum that sometimes goes by the name of the 'mind–body problem'. For thousands of years a popular view of consciousness, or mind, has been that it is a universal basic feature of all things. Known as panpsychism, this doctrine had many variations, but the common feature is the belief that mind suffuses the cosmos as an elementary quality; human consciousness is just an expression, focused and amplified, of a universal mental essence. In this respect it has elements in common with vitalism. Such thinking persisted well into the twentieth century; aspects of it can be found in Jung's psychology, for example. However, panpsychism doesn't sit comfortably with modern neuroscience, which emphasizes electrochemical complexity. In particular, higher brain functions are clearly associated with the collective organization of the neural architecture. It would make little sense to say that every neuron is 'a little bit' conscious and thus a collection of many neurons is very conscious. Only when millions of neurons are integrated into a complex and highly interconnected network does consciousness emerge. In the human brain, a conscious experience is made up of many components present simultaneously. If I am conscious of, say, a landscape, the momentary experience of the scene includes visual and auditory information from across the field of view, elaborately processed in different regions of the brain, then integrated into a coherent whole and (somehow!) delivered to 'the conscious self' (whatever that is) as a meaningful holistic experience.

All of which prompts the curious question, *where precisely* are minds? The obvious answer is: somewhere between our ears. But again, we can't be completely sure. For a long while the source of feelings was associated not with the brain but with other organs, like the gut, heart and spleen. Indeed, a vestige of this ancient belief lives on when angry people are described as 'venting their spleen' or we refer to a 'gut feeling' to mean intuition. And use of terms like 'sweetheart', 'heartthrob' and 'heartbroken' in the matter of romantic love are very common. It's unlikely that the endearment 'you are my sweetbrain' (still less 'my sweetamygdala') would serve to 'win the heart' of a lady, even though it is scientifically more accurate.

More radically, how can we be sure that the source of consciousness lies within our bodies at all? You might think that because a blow to the head renders one unconscious, the 'seat of consciousness' must lie within the skull. But there is no logical reason to conclude that. An enraged blow to my TV set during an unsettling news programme may render the screen blank, but that doesn't mean the news reader is situated inside the television. A television is just a receiver: the real action is miles away in a studio. Could the brain be merely a receiver of 'consciousness signals' created somewhere else? In Antarctica, perhaps? (This isn't a serious suggestion – I'm just trying to make a point.) In fact, the notion that somebody or something 'out there' may 'put thoughts in our heads' is a pervasive one; Descartes himself raised this possibility by envisaging a mischievous demon messing with our minds. Today, many people believe in telepathy. So the basic idea that minds are delocalized is actually not so far-fetched. In fact, some distinguished scientists have flirted with the idea that not all that pops up in our minds originates in our heads. A popular, if rather mystical, idea is that flashes of mathematical inspiration can occur by the mathematician's mind somehow 'breaking through' into a Platonic realm of mathematical forms and relationships that not only lies beyond the brain but beyond space and time altogether.[5] The cosmologist Fred Hoyle once entertained an even bolder hypothesis: that quantum effects in the brain leave open the possibility of external input into our thought processes and thus guide us towards useful scientific concepts. He proposed that this 'external guide' might be a superintelligence in the far cosmic future using a subtle but well-known backwards-in-time property of quantum mechanics in order to steer scientific progress.[6] Even if such wild notions are dismissed, extended minds could become the norm in the future. Humans may enjoy enhanced intelligence by outsourcing some of their mental activity to powerful computational devices that might be located in the cloud and coupled to their brains via wi-fi, thus repurposing brains as part receivers and part producers of consciousness.

An extreme version of the conjecture that our thoughts are generated outside our brains is the simulation argument, currently fashionable among certain philosophers and popularized by movies like *The Matrix*. The general idea is that what we take to be 'the real world' is actually a fancy virtual-reality show created inside a super-duper computer in the really real world. In this scheme, we human beings are modules of simulated consciousness.* Nothing can be said about the simulators – who or what they are, or what *it* is – because we poor simulations are stuck inside the system and so can never access the transcendent world of the simulator/s. In our fake simulated world, we have (fake) simulated bodies that include simulated brains, but the actual thoughts, sensations, feelings, and so on, that go along with consciousness don't arise in the fake brains at all but in the simulating system in another plane of existence altogether.

It's fun to speculate about these outlandish scenarios, but from here on I'm going to stick to the conservative view that consciousness is indeed produced, somehow, in the brain and ask what sort of physical process can do that. Don't be disappointed with this narrow agenda: there are still plenty of challenging problems to grapple with.

MIND OVER MATTER

Even non-solipsists – those who accept that other humans are conscious – cannot agree about which *non*-human organisms are conscious. Most people seem to be comfortable with the assumption that their pets have minds, but sliding down the tree of life towards its primitive trunk reveals no sharp boundary, no behavioural clues

---

* 'Computer' is a poor choice of word here as what we today think of as a computer almost certainly couldn't simulate consciousness.

that 'there is something in there'. Is a mouse conscious? A fly? An ant? A bacterium? If we want to argue by analogy, an important feature of consciousness is awareness of surroundings and an ability to respond appropriately to changes. Well, bacteria move towards food with what seems like purposeful agency. Yet it's hard to imagine that a bacterium can really 'feel hungry' in the same manner as you or I. But who can say?

Sometimes appeal is made to brain anatomy. It is clear that most of what the brain and associated nervous system does is performed unconsciously. Basic housekeeping functions – sensory signal processing and integration, searching memory, controlling motor activity, keeping the heart beating – proceed without our being aware of it. Many regions of the brain tick over just fine when someone loses consciousness (for example, in deep sleep, or when anaesthetized), which suggests that not all the brain is conscious or, more precisely, that generating consciousness is a function confined to only part of the brain, often taken to be a region called the corticothalamus. But it is difficult to determine exactly what properties this region possesses that other, unconscious yet still stupendously complex parts of the brain do not possess. Furthermore, some animals that display intelligent behaviour, such as birds, have very differently organized brain anatomy, so either consciousness and intelligence don't go hand in hand or attributing consciousness to a particular brain region is misconceived.

One thing isn't contentious: the brain processes information. It is therefore tempting to seek 'the source of consciousness' in the patterns of information swirling inside our heads. Neuroscientists have made huge strides in mapping what is going on in the brain when the subject experiences this or that sensation, emotion or sensory input. It isn't easy. The human brain contains 100 billion neurons (about the same as the number of stars in the galaxy) and each neuron connects with hundreds, maybe thousands, of others to form a vast network of information flow. Billions of rapid-firing neurons send elaborate cascades of electrochemical signals coursing through the network. Somehow, out of this electrical melee coherent consciousness emerges.

Distilling the problem down to basics, what we would like to know are the answers to the following two questions:

1. *What sort of physical processes generate consciousness?* This was what Schrödinger asked. For example, swirling electrical patterns of the sort that occur in brains would seem to, but what about swirling electrical patterns in the national power grid? If you answer yes to the first example and no to the second, then the question arises of whether it's all down to the *patterns*, as opposed to the electricity as such. Is there a pattern complexity threshold, so that brains are complex enough but electricity grids aren't? And if it's patterns that count, must it be done with electricity, or would any complex shimmering pattern do? Turbulent fluids, perhaps? Or interlocking chemical cycles? Alternatively, could it be that some other ingredient is needed – what one might call the 'electricity plus' theory of consciousness? And if so, what is the 'plus' bit? Nobody knows.

2. *Given that minds exist, how are they able to make a difference in the physical world?* How do minds couple to matter to give them causal purchase over material things? This is the ancient mind–body problem. If I choose to move my arm and my arm moves, something in the physical universe has changed (the position of my arm). But how does that happen? How is 'choice' or 'decision' transduced into movement of atoms? It's no good telling me that my desire to move my arm is nothing but swirling electrical patterns which then trigger electrical signals that travel through the nerves to my arm and cause muscle contraction, because that just purports to explain mystery 2 by appealing to mystery 1.

Running through my description is a hidden assumption always implicit in discussions of consciousness, namely, that there exists an agent or person or entity that 'possesses' consciousness. A mind 'belongs' to someone. I'm referring of course to the sense of self. Strictly, we must differentiate between being conscious of the world and being conscious of oneself ('self-consciousness'); perhaps a fly is conscious of the world but not of its own existence as an agent. But humans undeniably have a deep sense of self,* of being some sort of 'ghost in a machine'. Whatever the philosophical shortcomings of such dualism, it seems safe to say that almost everyone regards minds as

---

* Psychology experiments with small children suggest that full self-awareness does not develop until the age of about two.

real. But what are they? Not material or etherial substances. Information, perhaps? Not just any old information, but very specific patterns of information swirling in the brain. The general notion that information flow in neural circuitry somehow generates consciousness seems obvious, but a full explanation for mind needs to go much further. If the informational basis of mind is right, then minds exist in the same sense that information exists. But we cannot disconnect mind from matter. As Rolf Landauer taught us, 'information is physical', so minds must perforce also be tied to the material goings-on in the brain.

But *how*?

THE FLOW OF TIME

'The past, present and future is only a stubbornly persistent illusion.'

– Albert Einstein[7]

One clue to the link between neural information and consciousness comes from the most elementary aspect of human experience: our sense of the flow of time. Even under sensory deprivation people retain a sense of self and their continuing existence, so time's passage is an integral part of self-awareness. In Chapter 2 I described the existence of an arrow of time that can be traced back to the second law of thermodynamics and, ultimately, to the initial conditions of the universe. There is no disagreement about that. However, many people conflate the physical arrow of time with the psychological sense of the *flow* of time. Popular-science articles commonly use phrases like 'time flowing forwards' or the possibility of 'time running backwards'.

It's obvious that everyday processes possess an inbuilt directionality in time, so that if we witnessed the reverse sequence – like eggs unbreaking and gases unmixing all by themselves – we would be flabbergasted. Note that I am careful to describe sequences of physical states *in* time, yet the standard way of discussing it is to refer to an arrow *of* time. This misnomer is seriously misleading. The arrow is not a property of *time itself*. In this respect, time is little different from space. Think of the spin of the Earth, which also defines an asymmetry (between north and south). We sometimes denote that by an arrow too: a compass needle points north, and on a map it is conventional to show an arrow pointing north. We would never dream of saying, however, that Earth's north–south asymmetry (or the arrow on a map) is an 'arrow of space'. Space cares nothing for the spinning Earth, or north and south. Similarly, time cares nothing for eggs breaking or reassembling, or gases mixing or unmixing.

To call the sensation of a temporal *flow* 'the arrow of time', as is so often done, clearly conflates two distinct metaphors. The first is the use of an arrow to indicate spatial orientation (as in a compass needle), and the second is by analogy with an arrow in flight, symbolizing directed motion. When the arrow on a compass needle points north it doesn't indicate that you are *moving* north. In the same way, it is fine to attach an arrow to sequences of events in the world in order to distinguish past from future in the sequence, but what is not fine – what is absurd, in fact – is to then say that this arrow of asymmetry implies a *movement* towards the future along the timeline of events, that is, a movement *of* time.

My argument is further strengthened by noting that the alleged passage of time can't be measured. *There is no laboratory instrument that can detect the flow of time.* Hold on, you might be thinking, don't clocks measure time's passage? No, actually. A clock measures *intervals* of time between events. It does this by correlating the positions of the clock hands with a state of the world (for example, the position of a ball, the mental state of an observer). Informal descriptions like 'gravity slows time' and 'time runs faster in space than on Earth' really mean that the hands of clocks in space rotate slower relative to the hands of identical clocks on Earth. (They do. It is easy to test by comparing clock readings.) The most abusive terminology of all is talk about 'time running backwards'. Time doesn't 'run' at all. A correct rendition of the physics here is the possible reversal in (unchanged) time of the normal directional sequence of physical states, for example, rubble spontaneously assembling itself into buildings during an earthquake, Maxwell demons creating order out of chaos. It is not time itself but *the sequence of states* which 'goes backwards'.

In any case, it's obvious that time can't move. Movement describes the change of state of something (for example, the position of a ball) from one time to a later time. Time itself cannot 'move' unless there was a second time dimension relative to which its motion could be judged. After all, what possible answer can there be to the question 'How fast does time pass?' It has to be, 'One second per second' – a tautology! If you are not convinced, then try to answer the question 'How would you know if the rate of passage of time changed?' What would be observably different about the world if time speeded up or slowed down? If you woke up tomorrow and the rate of the flow of time had doubled, along with the rate of your mental processes, then nothing would appear to have changed, for the same reason that if you woke up and everything in the world was twice as big but so were you, nothing would look any different. Conclusion: the 'flow of time' makes no sense as a literal flow.

Although the foregoing points have been made by philosophers for over a century, the flow-of-time metaphor is so powerful that normal discourse is very hard without lapsing into it. Hard, but not impossible. Every statement about the world that makes reference to the passage of time can be replaced by a more cumbersome statement that makes no reference whatever to time's passage but merely correlates states of the world at various moments to brain/mind states at those same moments. Consider for example the statement 'With great anticipation we watched enthralled as the sun set over the ocean at 6 p.m.' The same basic observable facts can be conveyed by the ungainly statement: 'The clock configuration 5.50 p.m. correlates with the sun above the horizon and the observers' brain/mental state being one of anticipation; the clock configuration 6.10 p.m. correlates with the sun being below the horizon and the observers' brain/mental state being one of enthralment.' Informal talk about flowing or passing time is indispensable for getting by in daily life but makes no sense when traced back to the physics of time itself.

It is incontestable that we possess a very strong *psychological* impression that our awareness is being swept along on an unstoppable current of time, and it is perfectly legitimate to seek a scientific explanation for the *feeling* that time passes. The explanation of this familiar psychological flux is, in my view, to be found in neuroscience, not in physics. A rough analogy is with dizziness. Twirl around a few times and suddenly stop: you will be left with a strong impression that the world is rotating about you, even though it obviously isn't. The phenomenon can be traced to processes in the inner ear and brain: the feeling of continuing rotation is an illusion. In the same way, the sense of the motion of time is an illusion, presumably connected in some way to the manner in which memories are laid down in the brain.

To conclude: time doesn't pass. (I hope the reader is now convinced!)

Well, what *does* pass, then? I shall argue that it is the conscious awareness of the fleeting self that changes from moment to moment. The misconception that time flows or passes can be traced back to the tacit assumption of a *conserved* self. It is natural for people to think that 'they' endure from moment to moment while the world changes because 'time flows'. But as Alice remarked in Lewis Carroll's story, 'It's no use going back to yesterday, because I was a different person then.'[8] Alice was right: 'you' are not the same today as you were yesterday. To be sure, there is a very strong correlation – a lot of mutual information, to get technical about it – between today's you and yesterday's you – a thread of information made up of memories and beliefs and desires and attitudes and other things that usually change only slowly, creating an impression of continuity. But continuity is not conservation. There are future yous correlated with (that is, observing) future states of the world, and past yous correlated with (observing) past states of the world. At each moment, the you appropriate to that world-state interprets the correlation with that state as 'now'. It is indeed 'now' for 'that you' at 'that time'. That's all!

The flow-of-time phenomenon reveals 'the self' as a slowly evolving complex pattern of stored information that can be accessed at later times and provide an informational template against which fresh perceptions can be matched. The illusion of temporal flow stems from the inevitable slight mismatches.

DEMONS IN THE WIRING

So much for the elusive self. What about the brain? Here we are on firmer ground. Even on rudimentary inspection, it is clear that the brain is a ferment of electrochemical activity. First, some mind-blowing statistics. Recall that the human brain has about 100 billion neurons. These brain cells are powerhouses of information processing.[9] Each has a fibre called an axon sprouting from its body; it can be as long as one metre or more. Axons serve as wires that link neurons together to form a network. And what a dense network it is. Each neuron can be connected to up to 10,000 others; axons can branch hundreds of times. An axon doesn't just patch straight into another cell. Instead, neurons are decorated with a dense thicket of hair, or dendrites, and the axons clamp on to one of them. Other axons can attach to other dendrites of the same neuron, offering the opportunity to combine the incoming signals from many axons at once. It has been estimated that there could be as many as 1,000 trillion connections in the human brain as a whole, amounting to an astonishing level of complexity. Neurons can 'fire' (send pulses down axons) at a frenetic rate – maybe fifty times a second. All that adds up to the brain executing about 10^{15} logical operations per second, faster than the world's fastest supercomputer. Most arresting of all is that a supercomputer generates many megawatts of heat, whereas the brain does all that work with the same thermal output as a single low-wattage light bulb! (Impressive though that may be, brains still operate many orders of magnitude above the Landauer limit – see p. 46.)

The brain is often compared to an electrical circuit, and it's correct that the flow of electricity underlies its operation. But whereas the electrical signals in a computer (or the power grid) consist of electrons flowing down wires, the analogue of the wires in the brain – the axons – operate very differently. All along the axon are tiny holes in its outer membrane that can be opened and closed to let through one particle at a time, very much like Maxwell's original conception of a demon operating a shutter.[10] In this case, specialized proteins select different ions – charged atoms – rather than molecules. The holes are in fact narrow tubes, called 'voltage gated ion channels'; they can open and close a gate to let the right ions through and shut out the wrong ones. The way in which this set-up creates an electrical signal propagating down the axon is as follows. When the neuron is inert, the axon has a negative charge inside and a positive charge outside, creating a small voltage, or polarity, across the membrane. The membrane itself is an insulator. In response to the arrival of a signal from the body of the neuron, the gates open and allow sodium ions to flow from the outside to the inside, thereby reversing the voltage. Next, a different set of ion channels open to allow *potassium* ions to flow the other way – from the inside to the outside – restoring the original voltage. The polarity reversal typically lasts for only a few thousandths of a second. This transient disturbance triggers the same process in an adjacent section of the axon's membrane, and that in turn sets off the next section, and so on. The signal thus ripples down the axon towards another neuron. So although neurons signal each other electrically, it takes place via a travelling wave of polarity and not via a flow of electrical current as such.

To achieve this feat, the proteins need an astounding level of discrimination. In particular, they need to tell the difference between sodium ions and potassium ions (potassium ions are very slightly bigger) so as to let only the correct one through in the respective direction. The proteins stick through the membrane and provide a passage from the inside to the outside of the axon via an interior channel. The channel has a narrow bottleneck that allows one ion at a time to traverse it. Electric fields crafted by electrically polarized proteins maximize the efficiency; very little work is needed to push the ions through, and typical currents are millions of ions per second when the channel is open. The sorting precision is very high: less than one in 1,000 ions of the wrong species gets through. To decide when to open and close their gates, the protein clusters have sensors that can detect changes in the membrane potential nearby as the polarity wave approaches.

The upshot of all this demonic activity is that pulses, or spikes, of electric current travel down axons in groups, or trains, until they reach another neuron (sometimes another axon), where they can cause either excitation or inhibition of its activity. Neurons are not just passive relays that hand the signal on to the next neuron in line. They possess an internal structure that plays a critical role in processing the signal. Specifically, the axons are separated from the dendrites to which they attach by gaps about 20 nanometres wide called synapses, across which the signal may jump if the circumstances are right. The gap, known as a 'synaptic cleft', is mostly bridged not by an electric current as such but by a large variety of molecules called neurotransmitters. Some, like serotonin and dopamine, are familiar; others less so. These molecules are released from tiny vesicles (like mini-cells enclosed by a membrane) and diffuse across the cleft, where they bind to receptors on the far side. As a result of this binding, electrical changes are initiated in the body of the target neuron. For example, in its resting state, the neuron will have a negative charge relative to the outside of about 70 millivolts, maintained by pumping out ions through the cell membrane. The binding of neurotransmitters can cause the membrane to let through ions (for example, sodium, potassium, chloride) to alter that voltage. If the voltage drops below a certain threshold (that is, the inside of the cell is less negative), then the neuron will fire, sending a pulse down its axon to other neurons, and so on. Some neurotransmitters trigger an *increase* in membrane voltage (giving the interior of the cell an increased negative potential), which inhibits firing. Because converging incoming signals from many neurons can be amalgamated, the system acts rather like a logic circuit, with the neuron being either on (firing) or off (quiescent), according to the state of combination of the incoming signals.

How about the wiring architecture itself? Many of the details remain unknown, but the neural circuitry isn't static; it changes according to the individual's experiences. New memories, for example, are embedded by actively reorganizing the wiring. Thus, a baby is not born with a fixed 'circuit diagram' hard-wired in place but with a dense thicket of interconnections that can be pruned as well as rearranged as part of the growing and learning process.

HOW TO BUILD A MIND METER

If consciousness is an emergent, collective product of an organized whole, how can it be viewed in terms of information? It makes perfect sense to say that each neuron processes a few bits of information and a bundle of many neurons processes much more, but treating information arithmetically – just a head-count of bits bundled together – is merely another form of panpsychism. It fails to address the all-important property that information from across an extended region of the brain becomes *integrated* into a whole. An attempt to define a type of 'integrated information' as a measure of consciousness has been made by Giulio Tononi and his co-workers at the University of Wisconsin in Madison. The central idea is to capture in precise mathematical terms the intuitive notion that, when it comes to the brain, the whole is greater than the sum of its parts.

The concept of integrated information is clearest when applied to networks. Imagine a black box with input and output terminals. Inside are some electronics, such as a network with logic elements (AND, OR, and so on) wired together. Viewed from the outside, it will usually not be possible to deduce the circuit layout simply by examining the cause–effect relationship between inputs and outputs, because functionally equivalent black boxes can be built from very different circuits. But if the box is opened, it's a different story. Suppose you use a pair of cutters to sever some wires in the network. Now rerun the system with all manner of inputs. If a few snips dramatically alter the outputs, the circuit can be described as highly integrated, whereas in a circuit with low integration the effect of some snips may make no difference at all. To take a trivial example, suppose the box contains two separate self-contained circuits, each with its own input and output terminals. There could be wires cross-linking the two circuits that are totally redundant on account of the fact that no signals are ever directed down them. These wires can be severed with impunity.*

Tononi and his colleagues specify a way to calculate the irreducible interconnectedness of a general circuit by examining all possible decompositions of the circuit into fragments and working out how much information would be lost as a result. Highly integrated circuits lose a lot of information from the surgery. The precise degree of integration calculated this way is denoted by the Greek letter Φ. According to Tononi, systems with a big value of Φ, like the brain, are (in some sense) 'more conscious' than systems with small Φ, such as a thermostat. I should say that the precise definition of Φ is very technical; I

---

* It seems fair to say that if a neural network becomes *disintegrated* – if you carry on snipping – it will stop thinking altogether. Giulio Tononi's formula for measuring consciousness adopts this basic idea. However, there may be other formulae that would work better.

won't get into it here.[11] Generally speaking, if the elements in the box constrain each other's activity a great deal, then Φ is large; this will be the case if there are a lot of feedback loops and substantial 'cross-talk' – information transfer via cross-links. But if the system involves an orderly one-way flow of information from input to output (a feed-forward system), then $\Phi = 0$: what from outside the black box may appear as a unitary system is in fact just a conjunction of independent processes. Biology favours integrated systems – the brain being the supreme example – because they are more economical in terms of elements and connections, and more flexible than functionally equivalent systems with a purely feed-forward architecture. Larissa Albantakis, a member of Tononi's group, points out that the appearance of *autonomy* in a living organism (or a robot) goes hand in hand with high Φ: 'Being a causally autonomous entity from the intrinsic perspective requires an integrated cause–effect structure; merely "processing" information does not suffice.'[12] And there are surprises in store. The researchers find that, using their definition of Φ as a measure of con-sciousness, 'some simple systems can be minimally conscious, some complicated systems can be unconscious, and two different systems can be functionally equivalent, yet one is conscious and the other one is not'.[13]

In case the reader is lost in the technicalities here, let me offer an analogy. Imagine a twenty-member committee charged with the con-fidential task of awarding the annual Smith Prize for Scientific Excellence. The input data to the committee is the list of nominees and supporting documents; the output is the name of the winner. To the public, the committee seems like a 'black box': nominations go in, a recommendation comes out ('the committee has decided'). But now look at it from the internal perspective. If the members are independ-ent and vote without consultation, the committee is not integrated: it has $\Phi = 0$. But suppose there are factions – one group favours positive discrimination, another thinks the prize has gone to too many chem-ists, and so on. Because of their group affiliations, these members constrain each other's decisions; there is a measure of integration rep-resented by the 'cross-links' within each faction. If, further, there is extensive discussion within the committee (lots of feedback and cross-talk) following which a unanimous decision is made, then Φ is maximized. In the case that one member of the committee is a designated stenographer who records the proceedings but is not involved in the discussions, the committee has a lower value of Φ because it is not fully integrated.

There are inevitably many unanswered questions arising from identifying integrated information with consciousness, not least the extent to which actual neural function resembles the activities of logic circuits. Although computer comparisons are commonplace, most neuroscientists do not regard the brain as a souped-up digital computer. To be sure, the brain processes information, but using very different principles from the PC on which I am typing this. It is not even clear that *digital* is the way to go. Many neural functions may operate more like an analog computer. Nevertheless, integrated information is a laudable attempt to get to grips with consciousness in a quantitative way and to provide a theoretical underpinning based on causality and information flow.

FREE WILL AND AGENCY

'And so I say,
The atoms must a little swerve at times.'
— *Titus Lucretius Carus*[14]

A familiar property of human consciousness is a sense of freedom – a feeling we have that the future is somehow open, enabling humans to determine their own destiny, bending the arc of history according to their wills. Freedom means you may stop reading this chapter if you want (I hope you don't). In short, humans behave as agents.

A century ago free will seemed to be on a collision course with science. Brains are made of atoms, and atoms gotta do what atoms gotta do, that is, obey the laws of physics. For minds to influence the future by changing the activity in our brains (and thereby dictating our actions), they would have to exert physical forces in such a way that, crudely speaking, a brain atom happily moving to the left suddenly swerves to the right. This conundrum has been known since antiquity and was dubbed 'the atomic swerve' by Lucretius. A fully deterministic, mechanistic universe has no room for free will; the future is completely determined by the state of the universe today, right down to brains, neurons and the brain's atoms themselves. If the world is a closed mechanical system, then invoking a physical role for mind looks like a lost cause because it would imply *over-determinism*.

That was the situation in the late nineteenth century. Then along came quantum mechanics, with its inherent uncertainty. An atom that is moving to the left can indeed swerve to the right, all on its own, in the right circumstances, due to quantum weirdness. In the 1930s it seemed as if quantum indeterminism might rescue human free will. However, it's not that simple. To get free will we don't really want indeterminism: we want our wills to *determine* our actions. So a more subtle idea was floated. Maybe consciousness can indirectly affect atoms by 'loading the quantum dice', so that, although atoms may have an inherent propensity to behave capriciously, a type of bias or nudge here and there might creep in. This would give the mind a portal into the physical world, allowing it to inveigle its way by stealth into the quantum interstices of the causal chain. Unfortunately, even the inveigling would still amount to a violation of the laws of quantum mechanics in a statistical sense. Quantum physics may accommodate uncertainty, but it doesn't imply anarchy. Quantum mechanics involves very precise probabilistic rules, amounting to the equivalent of 'fair dice'. Mind-loaded dice would violate the quantum rules.

So what else is on offer? Scientists and philosophers have long wrestled with the problem of trying to reconcile the existence of agency with the underlying behaviour of the atoms and molecules that make up an agent. The agent doesn't have to be anything as complicated as a human being with considered motives. It may be a bacterium homing in on food. There is still a disconnect between the purposive behaviour of the agent and the blind, purposeless activities of the agent's components. How does purpose, or, if that word scares you, goal-oriented behaviour, emerge from atoms and molecules that care nothing about goals?

Information theory may have the answer. The first thing to notice is that agents are not closed systems. The very phenomenon of agency involves responding to changes in the system's environment. Living organisms are of course coupled to their environment in many ways, as I have been at pains to point out in earlier chapters. But even non-living agents, such as robots, are programmed to gather information from their surroundings, process it and effect an appropriate physical response. A truly closed system could not act (in a unitary manner) as an agent. So this provides a loophole in the problem of over-determinism. There is room for parallel narratives, one at the atomic level and another at the agent level, without contradiction, *so long as the system is open.*

Consider how the human brain is compartmentalized into many regions (left and right hemispheres, the thalamus, the cortex, the amygdala, and so on). Within this overall structure not all neurons are the same. Instead they are organized into various modules and clusters according to their different functions. A loosely defined unit is a 'cortical column', a module consisting of several thousand neurons with similar properties which can be treated as a single population. For example, neuroscientists treat cortical columns as individual units when considering stimulation–response relationships. There is a well-mapped region of the brain corresponding to the sense of touch on the surface of the skin. Neurons hooked up to the thumb lie close to those for the index finger, for example. If someone pricks your thumb, a module of neurons 'lights up' in that specific region of your brain and may initiate a motor response (you might say, 'Ouch!'). A neuroscientist can give an account of this scenario in terms of cause and effect, involving the 'thumb module' as a kind of simplified uni-tary agent.

Everyone agrees that, as a practical matter, it is sensible to refer to higher-level modules in explanations of brain activity rather than resort to an inconceivably complicated description of every neuron. However, Tononi's integrated information theory shows that not only is a higher-level description simpler, but higher-level systems can actually process *more* information than their components. This counter-intuitive claim has been investigated by Erik Hoel, a former member of Tononi's research group now working at Columbia University. Hoel carried out a quite general mathematical analysis to investigate the effects of aggregating microscopic variables in some way (such as by black boxing – see p. 199), using something called 'effective information theory'.[15] He set out to find how agents, with their associated intentions and goal-oriented behaviour, can emerge from the underlying microscopic physics, which lacks those properties. His conclusion is that there can be causal relationships that exist *solely at the level of agents*. Counter to most reductionist thinking, the macroscopic states of a physical system (such as the psychological state of an agent) that ignore the small-scale internal specifics can actually have *greater* causal power than a more detailed, fine-grained description of the system, a result summed up by the dictum: 'macro can beat micro'.

In spite of these careful analyses, a hard-nosed reductionist may point out that *in principle* a complete description of the stimulus–response story will nevertheless be present at the atomic level of the system. But there is an obvious flaw in this tired old argument, because it fails to take into account the openness of 'the system'. Let me explain. Response times (to pricked thumbs, say) are typically of the order of one-tenth of a second. Now consider that the stimulus–response system may consist of thousands of neurons networked by millions of axons, with neurons firing at fifty times a second. Recall the discussion about the demonic regulation of sodium and potassium ions that enter and leave the axon to drive the propagation of the signal. A neuron firing at fifty times a second will send a signal down an axon that entails the exchange of millions of ions. So, during the tenth of a second that the thumb drama plays out, 'the system' will exchange trillions of atomic particles with the extra-neuronal environment. The exiting particles quit the organized causal chain of the system to be lost amid the random thermal noise of the milieu and replaced by others that swarm in. It thus makes no sense to try to locate the bottom level of information about the thumb-pricking episode at the atomic scale. *Even in principle*, the cause–effect chain we are trying to explain simply *does not exist* at that level.

Well, counters the die-hard reductionist: what if one takes into account the environment of the system too, and the environment of that system, and so on, until our purview encompasses the entire cosmos? In principle (the argument goes), everything that happens, including the activity of brain modules, could then be accounted for at the atomic or subatomic level. Thus (says the reductionist), invoking the openness of an agent in order to rescue free will is to appeal to a pseudo-loophole. In my opinion, however, the reductionist's argument (which is often made by distinguished scientists) is absurd. There is no evidence that the universe is a closed deterministic system; it could be infinite. And even if it isn't, it's an indeterministic quantum system anyway.

QUANTUM BRAINS

Although quantum indeterminism can't explain deterministic wills, the perceived link between quantum mechanics and the mind is nevertheless deep and enduring. The nexus between the shadowy quantum domain and the world of concrete daily experience is an arena where one might expect mind and matter to meet. In quantum physics this is referred to as the 'measurement problem'. Here is why it is a problem. I explained in Chapter 5 how, at the atomic level, things get weird and fuzzy. When a quantum measurement is made, however, the results are sharp and well defined. For example, if the position of a particle is measured, a definite result is obtained. So what was previously fuzzy is suddenly focused, uncertainty is replaced by certainty, many contending realities are replaced by a single specific world. The difficulty now arises that the measuring system, which may consist of a piece of apparatus, a laboratory, a physicist, some students, and so on, is itself made of atoms subject to quantum rules. And there is nothing in the rules of quantum mechanics as formulated by Schrödinger and others to project out a particular, single, concrete reality from the legion of ghostly overlapping pseudo-realities characteristic of the quantum micro-world. So vexatious is this problem that a handful of physicists, including John von Neumann of universal constructor fame, suggested that the 'concretizing factor' (often called 'the collapse of the wave function') might be the *mind* of the experimenter. In other words, when the result of the measurement enters the consciousness of the measurer – *wham!* – the nebulous quantum world out there abruptly gels into commonsense reality. And if mind can do *that*, surely it does have a kind of leverage over matter, albeit in a subtle manner? It has to be admitted that today there are only a few adherents of this mentalist interpretation of quantum measurement, although there is still no consensus on a better explanation of just what happens when a quantum measurement takes place.

A new twist in the relationship between quantum fuzziness and human consciousness was introduced about thirty years ago by the Oxford mathematician Roger Penrose.[16] If consciousness somehow influences the quantum world, then, by symmetry, one might expect quantum effects to play a role in generating consciousness, and it's hard to see how that could happen unless there are quantum effects in the brain. In Chapter 5 I described how the field of quantum biology might explain photosynthesis and bird navigation, so a priori it seems not unreasonable that the behaviour of neurons might be influenced by quantum processes too. And that's what Penrose suggests is the case. More precisely, he claims that some microtubules threading through the interior of neurons might process information quantum mechanically, thus greatly boosting the processing power of the neural system and, somehow, generating consciousness on the way.[17] In arriving at this conclusion, Penrose and his colleague, the anaesthesiologist Stuart Hameroff, took into account the effects of anaesthesia, which occurs when a variety of molecules seeping into the neuronal synapses eliminate consciousness while leaving much of the routine functions of the brain unaffected – a process still not fully understood.

It has to be said that the Penrose–Hameroff theory has attracted a great deal of scepticism. Objections hinge on the problem of decoherence, which I explained in Box 11. Simple considerations imply that, in the warm and noisy environment of the brain, quantum effects would decohere very much faster than the speed of thought. Nevertheless, precise conclusions are hard to come by, and quantum mechanics has sprung surprises before.

I earlier described how Giulio Tononi and his colleagues have defined a quantity called integrated information, denoted Φ, which they offer as a mathematical measure of the degree of consciousness. Their ideas provide another way to link quantum mechanics to consciousness. Recall that integrated information quantifies the extent to which the whole may be greater than the sum of its parts when a system is complex. It thus depends on the state of the system as a whole – not just on its size or complexity but on the organization of its components and their relationship to the totality. A simple quantum system like an atom has a very low Φ but, if the atom is coupled to a measuring device, then the Φ of the whole system might be large, depending on the nature of the device. It would certainly be very large if a conscious human being were included in the system as part of the 'device', but the human element is not necessary. What if the way the quantum system changes in time depends on the value of Φ? Then, left alone, the atom would simply obey the normal rules of quantum physics applied to atoms that were presented by Schrödinger in the 1920s. But for a sufficiently complex system with significant integrated information (for example, a human observer), Φ would become important, eventually bringing about the wave function's collapse – that is, projection into a single concrete reality. What I am proposing is another example of top-down causation,[18] where the system as a whole (in this case precisely defined in terms of integrated information) exercises causal purchase over a lower-level component (the atom). In my example it is top-down causation defined in terms of *information* and so provides a clear example of an informational law entering into fundamental physics.[19]

Whatever the merits of these speculative ideas, I think it fair to say that if consciousness is ever to be fitted into the framework of physical theory, then it needs to be incorporated in *some* fashion into quantum mechanics, because quantum mechanics is our most powerful description of nature. Either consciousness violates quantum mechanics or it is explained by it.

Consciousness is the number-one problem of science, of existence even. Most scientists just steer clear of it, thinking it too much of a quagmire. Those scientists and philosophers who have dived in have usually become stuck. Information theory offers one way forward. The brain is an information-processing organ of stupendous complexity and intricate organization. Looking back at the history of life, each major transition has involved a reorganization of the informational architecture of organisms; the brain is the most recent step, creating information patterns that think.

Not everyone agrees, however, that cracking the information architecture problem will 'explain' consciousness, even if one buys into the thesis that conscious experiences are all about information patterns in the brain. David Chalmers, an Australian philosopher at New York University, divides the topic into 'the easy problem' and 'the hard problem'.[20] The easy part – very far from easy in practice – is to map the neural correlates of this or that experience, that is, determine which bit of the brain 'lights up' when the subject sees this or hears that. It's a doable programme. But knowing all the correlates still wouldn't tell us 'what it is like' to have this or that experience. I'm referring to the inner subjective aspect – the redness of red, for example – what philosophers call 'qualia'. Some people think the hard problem of qualia can never be settled, partly for the same reason that I can't be sure that you exist just because you behave more or less like I do. If so, the question 'What is mind?' will lie forever beyond our ken.

Epilogue

'One can best feel in dealing with living things how primitive physics still is.'

— *Albert Einstein*[1]

When Schrödinger delivered his Dublin lectures in 1943 he threw down a challenge that still resonates today. Can life be explained in terms of physics or will it always be a mystery? And if physics can explain life, is existing physics up to the job, or might it require something fundamentally new – new concepts, new laws even?

In the past few years it has become increasingly clear that information forms a powerful bridge between physics and biology. Only very recently has the interplay of information, energy and entropy been clarified, a century and a half after Maxwell introduced his notorious demon. Advances in nanotechnology have enabled incredibly delicate experiments to be performed to test foundational issues at the intersection of physics, chemistry, biology and computing. Though these developments have provided useful clues, so far the application of the physics of information to living systems has been piecemeal and ad hoc. Still lacking is a comprehensive set of principles that will explain all the puzzles in the magic box of life within a unitary theory.

While it is the case that biological information is *instantiated* in matter, it is not *inherent* in matter. Bits of information chart their own course inside living things. In so doing, they don't violate the laws of physics, but nor are they encapsulated by those laws: it is impossible to derive the laws of information from the known laws of physics. To properly incorporate living matter into physics requires new physics. Given that the conceptual gulf between physics and biology is so deep, and that existing laws of physics already provide a perfectly satisfactory explanation of the *individual* atoms and molecules that make up living organisms, it is clear that a full explanation of living matter entails something altogether more profound: nothing less than a revision of the nature of physical law itself.

Physicists have traditionally clung to a very restrictive notion of laws, dating from the time of Newton. Physics as we know it developed in seventeenth-century Europe, which was in thrall to Catholic Church doctrine. Although Galileo, Newton and their contemporaries were influenced by Greek thought, their notion of physical laws owed much to monotheism, according to which an omnipotent deity ordered the universe in a rational and intelligible manner. Early scientists regarded the laws of physics as thoughts in the mind of God. Classical Christian theology held that God is a perfect, eternal, unchanging being, transcending space and time. God made a physical world that changes with time, but God remains immutable. Creator and creature are thus not in a symmetrical relationship: the world depends utterly on God for its continued existence, but God does not depend on the world. Since it was held that the laws of the universe reflect the divine nature, it followed that the laws must also be unchanging. In 1630 Descartes expressed this very point explicitly:

> It is God who has established the laws of nature, as a King establishes laws in his kingdom ... You will be told that if God has established these truths, he could also change them as a King changes his laws. To which it must be replied: yes, if his will can change. But I understand them as eternal and immutable. And I judge the same of God.[2]

For these essentially theological reasons, physics was founded three centuries ago with a corresponding asymmetry between fixed laws and a changing world. That idea has been around so long we scarcely notice what a *huge* assumption it is. But there is no *logical* requirement it must be so, no compelling argument why the laws themselves have to be fixed absolutely. Indeed, I have already discussed one well-known example from fundamental physics in which the laws *do* change according to circumstance: the act of measurement in quantum mechanics. Measuring or observing a quantum system brings about a dramatic change in its behaviour, often called 'the collapse of the wave function'. To recap, it goes like this. Left alone, a quantum system (for example, an atom) evolves* according to a precise mathematical law provided by Schrödinger. But when the system is coupled to a measuring device and a measurement of a quantity is performed – for example, the energy of an atom – the state of the atom suddenly jumps ('collapses'). Significantly, the former evolution is reversible, but the latter is irreversible. So there are two completely different types of law for quantum systems: one when they are left alone and another when they are probed. Note a clue here linking to information. By performing a measurement of a quantum system the experimenter gains information about it (for example, which energy level an atom is in), but the entropy of the measured system jumps: we know less about its prior state after the measurement than we did before because of the irreversible 'collapse'.† So something has been gained and something lost.

Turning to biology, it is obvious that the notion of immutable laws is not a good fit. Darwin himself stressed the difference long ago in the closing passage of *On the Origin of Species*: '. . . whilst this planet has gone cycling on according to the fixed law of gravity, from so simple a beginning endless forms most beautiful and most wonderful have been, and are being, evolved'.[3] Biological evolution, with its open-ended variety and novelty and its lack of predictability, stands in stark contrast to the way that non-living systems evolve. Yet biology is not chaos: there are many examples of 'rules' at work, but these rules mostly refer to the *informational* architecture of organisms. Take the genetic code: the triplet of nucleotides CGT, for example, codes for the amino acid arginine (see Table 1). Although there are no known exceptions to that rule, it would be wrong to think of it as a law of nature, like the fixed law of gravity. Almost certainly the CGT $\rightarrow$ arginine assignment emerged a long time ago, probably from some earlier and simpler rule. Biology is full of cases like this; some rules are widespread, like Mendel's laws of genetics,

---

* The word 'evolve' has a very different meaning in physics from the one in use in biology, which can cause confusion.

† Irreversibility arises because the information about the phases of the various branches of the wave function has been destroyed by the act of measurement.

others more restrictive. When we consider the great drama of evolutionary history, the game of life must be seen as a game of quasi-rules that change over time.

More relevant is that the rules often depend on the *state* of the system concerned. To make this crucial point clear, let me give an analogy. Chess is a game with fixed rules. The rules don't determine the outcome of the game, the players do. There are a vast number of possible games, but a close inspection of all games would reveal that the pieces move across the board in accordance with the same rules. Now imagine a different type of chess game – call it chess-plus – in which the rules can change as the game progresses. In particular – to pursue the analogy with living systems – the rules could change depending on *the state of play*. One example might be this: 'if white is winning, then black is henceforth permitted to move the king up to two squares instead of one'. Here's another: 'if black has two more pawns than white, then white can move pawns backwards as well as forwards'. (These are silly suggestions, but some less drastic examples might pass muster as a popular game. Playing chess-plus, a novice might even beat a chess Grand Master.) The two examples I just gave involve 'rules of rule-change', or meta-rules, which are themselves fixed. But that's just for ease of exposition. The meta-rules don't have to be fixed: they could obey a meta-meta-rule or, to avoid an infinite regress, they could change randomly, perhaps decided by a coin toss. In the latter case chess-plus would become partly a game of skill and partly a game of chance. Either way, it is clear that chess-plus would be more complex and less predictable than conventional chess and would lead to states of play – that is, patterns of pieces on the board – that would be *impossible to attain* by following the conventional fixed rules of chess. We see here an echo of biology: life opens up regions of 'possibility space' that are inaccessible to non-living systems (see p. 10).

Laws that change as a function of the state are a generalization of the concept of self-reference: *what a system does depends on how a system is*. Recall from Chapter 3 that the notion of self-reference, following the work of Turing and von Neumann, lies at the core of both universal computation and replication. Relaxing the stringent requirement that laws have to be fixed and taking into account self-reference demands a whole new branch of science and mathematics, still largely unexplored. The physicist Nigel Goldenfeld of the University of Illinois is one of a handful of theorists who recognizes the promise of this approach: 'Self-reference should be an integral part of a proper understanding of evolution, but it is rarely considered explicitly,' he writes.[4] Goldenfeld contrasts biology with standard topics in physics like condensed matter theory, where 'there is a clear separation between the rules that govern the time evolution of the system and the state of the system itself ... the governing equation does not depend on the solution of the equation. In biology, however, the situation is different. The rules that govern the time evolution of the system are encoded in abstractions, the most obvious of which is the genome itself. As the system evolves in time, the genome itself can be altered, and so the governing rules are themselves changed. From a computer science perspective, one might say that the physical world can be thought of as being modelled by two distinct components: the program and the data. But in the biological world, the program is the data, and vice versa.'[5]

In Chapter 3 I described a simple attempt by my colleagues Alyssa Adams and Sara Walker to incorporate self-referential state-dependent rules in a cellular automaton (see p. 80). Sure enough, their computer model displayed the key property of open-ended variety that we associate with life. However, it was just a cartoon. To make the analysis realistic it would be necessary to apply self-referential state-dependent rules to information patterns in real complex physical systems. This hasn't been done – I'm throwing it out here as a challenge.[6] The resulting rules will differ from conventional laws of physics by applying at the *systems* level as opposed to individual components, such as particles, an example of top-down causation.[7] To be compatible with the laws of physics that we already know and love, any effects at the particle level would need to be small, or we would have noticed them already. But that is no obstacle. Because most molecular systems are inherently chaotic, inconspicuous, minute changes are able to accumulate and result in very profound effects. There is plenty of room at the bottom for novel physics to operate in a manner hitherto undetected and, indeed, that would be very hard to detect at the level of individual molecules anyway. But the cumulative impact on the information flow within an entire system, deriving from the combined effect of many tiny, disseminated influences, might come to dominate and yet appear inexplicable because the underlying causal mechanism has been overlooked.

The possibility that there may be new laws, or at least systematic regularities, hidden in the behaviour of complex systems, is by no means revolutionary. Several decades ago it was discovered that subtle mathematical patterns were buried in a wide range of *chaotic* systems ('chaotic' here means such systems are unpredictable even with a very precise knowledge of the forces and starting conditions, the weather being a classic example). Physicists began to talk about 'universality in chaos'. What I am proposing here is *universality in informational organization*, in the expectation that common information patterns will be found in a large class of certain complex systems – patterns that capture, at least in part, something of the features of living organisms.

So much for theory, which has barely scratched the surface of these new ideas. What are the prospects for experiment? Here we run up against the overwhelming complexity of biology. If the new informational state-dependent laws I am proposing operated only in living matter, it would be just another version of vitalism. The whole purpose of a theory that unifies physics and biology is to remove any barrier separating them, in which case the new informational laws might be expected to bleed from the living world into the non-living world. Several decades ago a claim to have discovered just such an effect was made by Sidney Fox, a biochemist based in Alabama who devoted his career to studying the origin of life. Fox published experimental evidence to suggest that when amino acids assemble into chains (called peptides), they show a preference for just those combinations that lead to biologically useful molecules, that is, proteins. 'Amino acids determine their own order in condensation,' he wrote.[8] If true, the claim would be evidence that the laws of chemistry somehow favoured life, as if they knew about it in advance. Even more dramatic were the claims of Gary Steinman and Marian Cole of Pennsylvania State University, who also reported non-random peptide formation: 'These results prompt the speculation that unique, biologically pertinent peptide sequences may have been produced prebiotically,' they wrote.[9]

The suggestion that chemistry is cunningly rigged in favour of life was widely dismissed, and indeed was scarcely credible in the form presented by Fox and others, involving as it did preferential bonding between pairs of molecules – a process well understood within the framework of quantum mechanics. But if one took an informational approach to molecular organization, it might be a different story.[10]

If we had properly worked-out candidates for informational state-dependent laws, they might suggest that systems self-organize in ways to amplify their information-processing abilities or lead to 'unreasonable' accumulation of integrated information. The recent discovery that in some circumstances 'macro beats micro' in terms of causal power (see p. 204) opens the possibility that the spontaneous organization of higher-order information-processing modules might be favoured as a general trend in complex systems. The pathway from non-life to life might be far shorter when viewed in terms of the organization of information rather than chemical complexity. If so, it would greatly boost the search for a second genesis of life.*

In this book I have charted a burgeoning new area of science. As I write, scarcely a day passes without the publication of another paper or the announcement of a new experimental result having a direct impact on the physics of information and its role in the story of life. This is a field in its infancy and many questions remain unanswered. If there are new physical laws at work – informational laws, perhaps involving state dependence and top-down causation – how do we mesh them with the known laws of physics? And would these new laws be deterministic in form or contain an element of chance, like quantum mechanics? Indeed, does quantum mechanics come into them? Does it in fact play an integral role in life? In addition to these imponderables lies the question of origins. How do life's informational patterns come into existence in the first place? The appearance of anything new in the universe is always an amalgam of

---

* Many scientists (de Duve included) support the idea of a cosmic imperative without feeling the need for novel laws or principles to fast-track life's genesis. They appeal to the generality of the laws of known chemistry and eschew parochialism – why should we/Earth be so special? But this has an air of wishful thinking to it. I am totally sceptical that known chemistry embeds a life principle, for known chemistry offers no conceptual bridge between molecules and information.

laws and initial conditions. We simply don't know the conditions nec-essary for biological information to emerge initially, or, once left to get going, how strong a role natural selection plays versus the oper-ation of informational laws or other organizational principles that may be at work in complex systems. All this has to be worked out.

There will be those who object to dignifying the informational principles I have been elucidating with the word 'law' in any deep sense. While most scientists are happy to treat information patterns as things in their own right for *practical* purposes, reductionists insist that this is merely a methodological convenience and that, in princi-ple, all such 'things' can be reduced to fundamental particles and the laws of physics – and hence defined out of existence. They don't 'really exist', we are warned, except in our own imaginings. While reductionists may concede that certain rules 'emerge' in complex sys-tems, they assert that these rules do not enjoy the fundamental status of the laws of physics that underlie all systems. The reductionist arg-ument is undeniably powerful, but it rests on a major assumption about the nature of physical law. The way the laws of physics are currently conceived leads to a stratification of physical systems with the laws of physics at the bottom conceptual level and emergent laws stacked above them. There is no coupling between levels. When it comes to living systems, this stratification is a poor fit because, in biology, there often is coupling between levels, between processes on many scales of size and complexity: causation can be both bottom-up (from genes to organisms) and top-down (from organisms to genes). To bring life within the scope of physical law – and to provide a sound basis for the reality of information as a fundamental entity in its own right – requires a radical reappraisal of the nature of physical law, as I am arguing. [11]

It would be wrong to think that these arcane deliberations are impor-tant only to a handful of scientists, philosophers and mathematicians. They have sweeping implications not just for explaining life but for the nature of human existence and our place in the universe. Before Dar-win, it was widely believed that God created life. Today, most people accept it had a naturalistic origin. While it is true that scientists lack a full explanation for how life emerged from non-life, invoking a one-off miracle is to fall into the god-of-the-gaps trap. It would imply a type of cosmic magician who sporadically intervenes, moving molecules around from time to time but mostly leaving them to obey fixed laws. Yet within the broad scope of the term 'naturalistic' lie very different philo-sophical (even theological) implications. Two contrasting views of life's origin are the statistical fluke hypothesis championed by Jacques Monod and the cosmic imperative of Christian de Duve. Monod appealed to the flukiness of life to bolster his nihilistic philosophy: 'The ancient cov-enant is in pieces,' he wrote gloomily. '[Man's] destiny is nowhere spelled out, nor is his duty. The kingdom above or the darkness below: it is for him to choose ... The universe was not pregnant with life, nor the biosphere with man.'[12] In responding to Monod's negative reflec-tions, de Duve wrote, 'You are wrong. They were,'[13] and proceeded to develop his view of what he called 'a meaningful universe'. Boiled down to basics, the issue is this. Is life built into the laws of physics? Do those laws magically embed the designs of organisms-to-be? There is no evi-dence whatever that the *known* laws of physics are rigged in favour of life; they are 'life-blind'. But what about new state-dependent infor-mational laws of the sort I am conjecturing here? My hunch is that they would not be so specific as to foreshadow biology as such, but they might favour a broader class of complex information-managing systems of which life as we know it would be a striking representative. It's an uplifting thought that the laws of the universe might be intrinsically bio-friendly in this general manner.

These speculative notions are very far from a miracle-working deity who conjures life into being from dust. But if the emergence of life, and perhaps mind, are etched into the underlying lawfulness of nature, it would bestow upon our existence as living, thinking beings a type of cosmic-level meaning.

It would be a universe in which we can truly feel at home.

Further Reading

1. WHAT *IS* LIFE?

Anthony Aguirre, Brendan Foster and Zeeya Merali (eds.), *Wandering towards a Goal: How Can Mindless Mathematical Laws Give Rise to Aims and Intention?* (Springer, 2018)

Philip Ball, 'How life (and death) spring from disorder', *Quanta*, 25 January 2017; https://www.quantamagazine.org/the-computational-foundation-of-life-20170126/

Steven Benner, *Life, the Universe and the Scientific Method* (The FfAME Press, 2009)

Paul Davies and Niels Gregersen (eds.), *Information and the Nature of Reality: From Physics to Metaphysics* (Cambridge University Press, 2010)

Nick Lane, *The Vital Question: Energy, Evolution and the Origins of Complex Life* (Norton, 2015)

Ilya Prigogine and Isabelle Stengers, *Order out of Chaos* (Heinemann, 1984)

Erwin Schrödinger, *What is Life?* (Cambridge University Press, 1944; Canto edn, 2012)

Sara Walker, Paul Davies and George Ellis (eds.), *From Matter to Life: Information and Causality* (Cambridge University Press, 2017)

Carl Woese, 'A new biology for a new century', *Microbiology and Molecular Biology Reviews*, vol. 68, no. 2, 173–86 (2004)

2. ENTER THE DEMON

Derek Abbott, 'Asymmetry and disorder: a decade of Parrondo's paradox', *Fluctuation and Noise Letters*, vol. 9, no. 1, 129–56 (2010)

R. Dean Astumian and Imre Derényi, 'Fluctuation driven transport and models of molecular motors and pumps', *European Biophysics Journal*, vol. 27, 474–89 (1998)

Peter Atkins, *The Laws of Thermodynamics: A Very Short Introduction* (Oxford University Press, 2010)

Philip Ball, 'Bacteria replicate close to the physical limit of efficiency', *Nature*, 20 September 2012; http://www.nature.com/news/bacteria-replicate-close-to-the-physical-limit-of-efficiency-1.11446

Charles H. Bennett, 'Notes on Landauer's principle, reversible computation and Maxwell's Demon', *Studies in History and Philosophy of Modern Physics*, vol. 34, 501–10 (2003)

Philippe M. Binder and Antoine Danchin, 'Life's demons: information and order in biology', *European Molecular Biology Organization (EMBO) Reports*, vol. 12, no. 6, 495–9 (2011)

S. Chen et al., 'Structural diversity of bacterial flagellar motors', *EMBO Journal*, 30 (14), 2972–81 (2011); doi: http://dx.doi.org/10.1038/emboj.2011.186

Kensaku Chida et al., 'Power generator driven by Maxwell's demon', *Nature Communications*, 8:15301 (2017)

Nathanaël Cottet et al., 'Observing a quantum Maxwell demon at work', *Proceedings of the National Academy of Sciences*, vol. 114, no. 29, 7561–4 (2017)

Alexander R. Dunn and Andrew Price, 'Energetics and forces in living cells', *Physics Today*, vol. 68, no. 2, 27–32 (2015)

George Dyson, *Turing's Cathedral: The Origins of the Digital Universe* (Vintage, 2012)

Lin Edwards, 'Maxwell's demon demonstration turns information into energy', *PhysOrg.com*, 15 November 2010; https://phys.org/news/2010-11-maxwell-demon-energy.html

Ian Ford, 'Maxwell's demon and the management of ignorance in stochastic thermodynamics', *Contemporary Physics*, vol. 57, no. 3, 309–30 (2016)

Jennifer Frazer, 'Bacterial motors come in a dizzying array of models', *Scientific American*, 16 December 2014

James Gleick, *The Information: A History, a Theory, a Flood* (HarperCollins, 2011)

Gregory P. Harmer et al., 'Brownian ratchets and Parrondo's games', *Chaos*, 11, 705 (2001); doi: 10.1063/1.1395623

Peter Hoffman, *Life's Ratchet* (Basic Books, 2012)

—, 'How molecular motors extract order from chaos', *Reports on Progress in Physics*, vol. 79, 032601 (2016)

William Lanouette and Bela Silard, *Genius in the Shadows: A Biography of Leo Szilárd, the Man behind the Bomb* (University of Chicago Press, 1994)

C. H. Lineweaver, P. C. W. Davies and M. Ruse (eds.), *Complexity and the Arrow of Time* (Cambridge University Press, 2013)

Norman MacRae, *John von Neumann: The Scientific Genius Who Pioneered the Modern Computer, Game Theory, Nuclear Deterrence, and Much More* (American Mathematical Society; 2nd edn, 1999)

J. P. S. Peterson et al., 'Experimental demonstration of information to energy conversion in a quantum system at the Landauer limit', *Proceedings of The Royal Society A*, vol. 472, issue 2188 (2016): 20150813

Takahiro Sagawa, 'Thermodynamic and logical reversibilities revisited', *Journal of Statistical Mechanics* (2014); doi: 10.1088/1742-5468/2014/03/P03025

Jimmy Soni and Rob Goodman, *A Mind at Play: How Claude Shannon Invented the Information Age* (Simon and Schuster, 2017)

3. THE LOGIC OF LIFE

Gérard Battail, *Information and Life* (Springer, 2014)

Gregory Chaitin, *The Unknowable: Discrete Mathematics and Theoretical Computer Science* (Springer, 1999)

Peter Csermely, 'The wisdom of networks: a general adaptation and learning mechanism of complex systems', *BioEssays*, 1700150 (2017)

Deborah Gordon, *Ants at Work: How an Insect Society is Organized* (Free Press, 2011)

Andrew Hodges, *Alan Turing: The Enigma: The Book that Inspired the Film 'The Imitation Game'* (Princeton University Press, 2014)

Douglas Hofstadter, *Gödel, Escher, Bach: An Eternal Golden Braid* (Basic Books, 1979)

Bernd-Olaf Küppers, *Information and the Origin of Life* (MIT Press, 1990)

Janna Levin, *A Madman Dreams of Turing Machines* (Knopf, 2006)

G. Longo et al., 'Is information a proper observable for biological organization?', *Progress in Biophysics and Molecular Biology*, vol. 109, 108–14 (2012)

Denis Noble, *Dance to the Tune of Life: Biological Relativity* (Cambridge University Press, 2017)

Paul Rendell, *Turing Machine Universality of the Game of Life: Emergence, Complexity and Computation* (Springer, 2015)

Stephen Wolfram, *A New Kind of Science* (Wolfram Media, 2002)

Hubert Yockey, *Information Theory, Evolution and the Origin of Life* (Cambridge University Press, 2005)

4. DARWINISM 2.0

Nessa Carey, *The Epigenetics Revolution: How Modern Biology is Rewriting Our Understanding of Genetics, Disease and Inheritance* (Columbia University Press, 2013)

Richard Dawkins, *The Selfish Gene* (Oxford University Press, 1976)

Daniel Dennett, *Darwin's Dangerous Idea: Evolution and the Meaning of Life* (Simon and Schuster, 1995)

Robin Hesketh, *Introduction to Cancer Biology* (Cambridge University Press, 2013)

Eva Jablonka and Marion Lamb, *Evolution in Four Dimensions* (MIT Press, 2005)

George Johnson, *The Cancer Chronicles: Unlocking Medicine's Deepest Mystery* (Vintage, 2014)

Stuart Kauffman, *The Origin of Order: Self-organization and Selection in Evolution* (Oxford University Press, 1993)

Lewis J. Kleinsmith, *Principles of Cancer Biology* (Pearson, 2005)

Matthew Niteki (ed.), *Evolutionary Innovations* (University of Chicago Press, 1990)

Massimo Pigliucci and Gerd B. Müller (eds.), *Evolution, the Extended Synthesis* (MIT Press, 2010)

Trygve Tollefsbol (ed.), *Handbook of Epigenetics* (Academic Press, 2011)

Andreas Wagner, *Arrival of the Fittest* (Current, 2014)

Robert A. Weinberg, *The Biology of Cancer* (Garland Science, 2007)

Edward Wilson, *The Meaning of Human Existence* (Liveright, 2015)

5. SPOOKY LIFE AND QUANTUM DEMONS

Derek Abbott, Paul Davies and Arun Patti (eds.), *Quantum Aspects of Life* (Imperial College Press, 2008)

Richard Feynman, 'Simulating physics with computers', *International Journal of Theoretical Physics*, vol. 21, nos. 6/7 (1982)

Johnjoe McFadden and Jim Al-Khalili, *Life on the Edge: The Coming of Age of Quantum Biology* (Bantam Press, 2014)

Masoud Mohseni, Yasser Omar, Gregory S. Engel and Martin B. Plenio (eds.), *Quantum Effects in Biology* (Cambridge University Press, 2014)

Leonard Susskind and Art Friedman, *Quantum Mechanics: The Theoretical Minimum* (Basic Books, 2015)

Peter G. Wolynes, 'Some quantum weirdness in physiology', *Proceedings of the National Academy of Sciences*, vol. 106, no. 41, 17247–8 (13 October 2009)

6. ALMOST A MIRACLE

A. G. Cairns-Smith, *Seven Clues to the Origin of Life: A Scientific Detective Story* (Cambridge University Press, 1985)

Matthew Cobb, *Life's Greatest Secret: The Race to Crack the Genetic Code* (Basic Books, 2015)

Paul Davies, *The Fifth Miracle: The Search for the Origin of Life* (Allen Lane, 1998)

Christian de Duve, *Vital Dust: The Origin and Evolution of Life on Earth* (Basic Books, 1995)

Freeman Dyson, *Origins of Life* (Cambridge University Press; 2nd edn, 1999)

Pier Luigi Luisi, *The Emergence of Life: From Chemical Origins to Synthetic Biology* (Cambridge University Press; 2nd edn, 2016)

Eric Smith and Harold Morowitz, *The Origin and Nature of Life on Earth* (Cambridge University Press, 2016)

Woodruff T. Sullivan III and John A. Baross (eds.), *Planets and Life* (Cambridge University Press, 2007)

Sara Walker and George Cody, 'Re-conceptualizing the origins of life', *Philosophical Transactions of The Royal Society* (theme issue), vol. 375, issue 2109 (2017)

7. THE GHOST IN THE MACHINE

David Chalmers, *The Conscious Mind: In Search of a Fundamental Theory* (Oxford University Press; rev. edn, 1997)

Daniel Dennett, *Consciousness Explained* (Little, Brown, 1991)

George Ellis, *How Can Physics Underlie the Mind? Top-down Causation in the Human Context* (Springer, 2016)

Douglas R. Hofstadter and Daniel C. Dennett, *The Mind's I: Fantasies and Reflections on Self and Soul* (Basic Books, 2001)

Stuart Kauffman, *At Home in the Universe: The Search for the Laws of Self-Organization and Complexity* (Oxford University Press, 1996)

Arthur Koestler, *The Ghost in the Machine* (Hutchinson, 1967)

Nancey Murphy, George F. R. Ellis and Timothy O'Connor (eds.), *Downward Causation and the Neurobiology of Free Will* (Springer, 2009)

Roger Penrose, *The Emperor's New Mind: Concerning Computers, Minds and the Laws of Physics* (Oxford University Press, 1989)

Bruce Rosenblum and Fred Kuttner, *Quantum Enigma: Physics Encounters Consciousness* (Oxford University Press; 2nd edn, 2011)

Notes

1. WHAT *IS* LIFE?

1. Erwin Schrödinger, *What is Life?* (Cambridge University Press, 1944), p. 23.
2. Charles Darwin, *On the Origin of Species* (John Murray; 2nd edn, 1860), p. 490.
3. David Deutsch, *The Beginning of Infinity: Explanations that Transform the World* (Penguin, 2011), p. 1.
4. S. I. Walker, 'The descent of math', in: A. Aguirre, B. Foster and Z. Merali (eds.), *Trick of Truth: The Mysterious Connection between Physics and Mathematics* (Springer, 2016).
5. Richard Dawkins, *Climbing Mount Improbable* (Norton, 1996).
6. Eric Smith and Harold Morowitz, *The Origin and Nature of Life on Earth* (Cambridge University Press, 2016).
7. Bernd-Olaf Küppers, 'The nucleation of semantic information in prebiotic matter', in: E. Domingo and P. Schuster (eds.), *Quasispecies: From Theory to Experimental Systems. Current Topics in Microbiology and Immunology*, vol. 392, 23–42. See also Carlo Rovelli, 'Meaning and intentionality = information + evolution', in: A. Aguirre, B. Foster and Z. Merali (eds.), *Wandering Towards a Goal: How Can Mindless Mathematical Laws Give Rise to Aims and Intention?* (Springer, 2018), pp. 17–27.
8. Eric Smith, 'Chemical Carnot cycles, Landauer's principle and the thermodynamics of natural selection', Talk/Lecture, Bariloche Complex Systems Summer School (2008).

2. ENTER THE DEMON

1. Peter Hoffman, *Life's Ratchet* (Basic Books, 2012), p. 136.

2. Claude Shannon, *The Mathematical Theory of Communication* (University of Illinois Press, 1949).

3. Christoph Adami, 'What is information?', *Philosophical Transactions of The Royal Society*, A 374: 20150230 (2016).

4. Leo Szilárd, 'On the decrease of entropy in a thermodynamic system by the intervention of intelligent beings', *Zeitschrift fur Physik*, 53, 840–56 (1929).

5. The term 'ultimate laptop' was coined by Seth Lloyd in an analysis of the most efficient possible computer in the universe. See https://www.edge. org/conversation/seth_lloyd-how-fast-how-small-and-how-powerful.

6. There remain some subtleties in this matter. It has been argued that there may be ways to circumvent Landauer's limit in special situations. See O. J. E. Maroney, 'Generalizing Landauer's principle', *Physical Review*, E 79, 031105 (2009). In some cases, logical irreversibility does not go hand in hand with thermodynamic irreversibility.

7. Rolf Landauer, 'Irreversibility and heat generation in the computing process', *IBM Journal of Research and Development*, 5 (3): 183–91; doi: 10.1147/rd.53.0183 (1961).

8. Charles Bennett and Rolf Landauer, 'The fundamental physics limits of computation', *Scientific American*, vol. 253, issue 1, 48–56 (July 1985).

9. Alexander Boyd and James Crutchfield, 'Maxwell demon dynamics: deterministic chaos, the Szilárd map, and the intelligence of thermodynamic systems', *Physical Review Letters*, 116, 190601 (2016).

10. Z. Lu, D. Mandal and C. Jarzynski, 'Engineering Maxwell's demon', *Physics Today*, vol. 67, no. 8, 60–61 (2014).

11. https://www.youtube.com/watch?v=ooTyIShzR6o

12. Lu, Mandal and Jarzynski, 'Engineering Maxwell's demon'.

13. Ibid.

14. Douglas Adams, *The Hitchhiker's Guide to the Galaxy* (Del Ray, 1995).

15. Katharine Sanderson, 'A demon of a device', *Nature*, 31 January 2007; doi: 10.1038/news070129-10.

16. Viviana Serreli et al., 'A molecular information ratchet', *Nature*, vol. 445, 523–7 (2007).

17. Quoted in Stephen Battersby, 'Summon a "demon" to turn information into energy', *New Scientist Daily News*, 15 November 2010. More recently, a Korean research group managed to convert information into work with an astonishing 98.5 per cent efficiency. See Lisa Zyga, 'Information engine operates with nearly perfect efficiency', Phys.Org.com, 19 January 2018.

18. J. V. Koski et al., 'On-chip Maxwell's demon as an information-powered refrigerator', *Physical Review Letters*, 115, 260602 (2015).

19. Christoph Adami, as quoted in 'The Information Theory of Life', by Kevin Hartnett, *Quanta* (19 November, 2015).

20. Kazuhiko Kinosita, Ryohei Yasuda and Hiroyuki Noji, 'F1-ATPase: a highly efficient rotary ATP machine', *Essays in Biochemistry*, vol. 35, 3–18 (2000).

21. https://www.youtube.com/watch?v=y-uuk4Pr2i8.

22. See account in, for example, Peter Hoffman *Life's Ratchet* (Basic Books, 2012), pp. 159–62.

23. Anita Goel, R. Dean Astumian and Dudley Herschbach, 'Tuning and switching a DNA polymerase motor with mechanical tension', *Proceedings of the National Academy of Sciences*, vol. 100, no. 17, 9699–704 (2003).

24. Jeremy England, 'Statistical physics of self-replication', *Journal of Chemical Physics*, vol. 139, 121923, 1–8 (2013).

25. Rob Phillips and Stephen Quake, 'The biological frontier of physics', *Physics Today*, vol. 59, 38–43 (May 2006).

26. For Feynman's own account, see Lecture 46 of his Caltech series: http://www.feynmanlectures.caltech.edu.

27. Andreas Wagner, 'From bit to it: how a complex metabolic network transforms information into living matter', *BMC Systems Biology*; doi: 10.1186/1752-0509-1-33 (2007).

28. This includes Bayesian inference. See David Spivak and Matt Thomson, 'Environmental statistics and optimal regulation', *PLoS Computational Biology*, vol. 10, no. 10 e 1003978 (2014).

3. THE LOGIC OF LIFE

1. 'What is life?: an interview with Gregory Chaitin', *Admin*: http://www.philosophytogo.org/wordpress/?p=1868 (18 December 2010).

2. Alan Turing, 'On computable numbers, with an application to the Entscheidungsproblem', *Proceedings of the London Mathematical Society*, ser. 2, vol. 42 (1937). See also http://www.turingarchive.org/browse.php/b/12.

3. Ibid.

4. John von Neumann, *Theory of Self-reproducing Automata* (University of Illinois Press, 1966).

5. George F. R. Ellis, Denis Noble and Timothy O'Connor, 'Top-down causation: an integrating theme within and across the sciences?', *Royal Society Interface Focus* (2012).

6. John L. Casti, 'Chaos, Gödel and truth', in: J. L. Casti and A. Karlqvist (eds.), *Beyond Belief: Randomness, Prediction and Explanation in Science* (CRC Press, 1991); M. Prokopenko et al., 'Self-referential

basis of undecidable dynamics: from the liar paradox and the halting problem to the edge of chaos', arXiv:1711.02456 (2017). Note that if the system is a finite state machine it will obviously generate only finite novelty. An infinite array would not have this restriction.

7. J. T. Lizier and M. Prokopenko, 'Differentiating information transfer and causal effect', *European Physical Journal B*, vol. 73, no. 4, 605–15 (2010); doi: 10.1140/epjb/e2010-00034-5.

8. Alyssa Adams at al., 'Formal definitions of unbounded evolution and innovation reveal universal mechanisms for open-ended evolution in dynamical systems', *Scientific Reports (Nature)*, vol. 7, 997–1012 (2017).

9. Richard Dawkins, *The Selfish Gene* (Oxford University Press, 1976).

10. Y. Lazebnik, 'Can a biologist fix a radio? Or, what I learned while studying apoptosis', *Biochemistry* (Moscow), vol. 69, no. 12, 1403–6 (2004).

11. Paul Nurse, 'Life, logic and information', *Nature*, vol. 254, 424–6 (2008).

12. Uri Alon, *An Introduction to Systems Biology: Design Principles of Biological Circuits* (Chapman and Hall, 2006).

13. Some people express puzzlement that a system as complicated as a human being could be the product of a mere 20,000 genes. Is there enough information in 20,000 genes? No, there isn't. But given that a gene can be in one of two states, there are in theory $2^{20,000}$ (about $10^{6,000}$) possible combinations of gene expression, which is vastly greater than all the bits of information in the universe (a paltry 10^{123}), let alone in a human being. Viewed this way, there is now an overabundance of available bits to go round.

14. Alon, *An Introduction to Systems Biology*.

15. Ibid.

16. Benjamin H. Weinberg et al., 'Large-scale design of robust genetic circuits with multiple inputs and outputs for mammalian cells', *Nature Biotechnology*, vol. 35, 453–62 (2017).

17. Hideki Kobayashi et al., 'Programmable cells: interfacing natural and engineered gene networks', *Proceedings of the National Academy of Sciences*, 8414–19; doi: 10.1073/pnas.0402940101 (2017).

18. There has been significant interest in encouraging education and outreach as well: the International Genetically Engineered Machines Competition allows undergraduate and high-school students to design their own synthetic biological circuits.

19. Maria I. Davidich and Stefan Bornholdt, 'Boolean network model predicts cell cycle sequence of fission yeast', *PLoS ONE*, 27 February 2008: https://doi.org/10.1371/journal.pone.0001672.

20. Hyunju Kim, Paul Davies and Sara Imari Walker, 'New scaling relation for information transfer in biological networks', *Journal of the Royal Society Interface* 12 (113), 20150944 (2015); doi: 10.1098/rsif.2015.0944.

21. Richard Feynman and Ralph Leighton, *Surely You're Joking, Mr. Feynman!* (Norton, 1985). See also https://www.youtube.com/watch?v=nmEoL5C7ths.

22. Uzi Harush and Baruch Barzel, 'Dynamic patterns of information flow in complex networks', *Nature Communications* (2017); doi: 10.1038/s41467-017-01916-3.

23. Nurse, 'Life, logic and information'.

4. DARWINISM 2.0

1. Theodosius Dobzhansky, 'Nothing in biology makes sense except in the light of evolution', *American Biology Teacher*, 35 (3): 125–9 (March 1973).

2. For a detailed popular account of how evolution selects for demonic information-processing efficiency see *The Touchstone of Life* by Werner Loewenstein (Oxford University Press, 1999), Ch. 6.

3. Eva Jablonka, *Evolution in Four Dimensions* (MIT Press, 2005), p. 1.

4. https://www.theregister.co.uk/2017/06/14/flatworm_sent_to_space_returns_2_headed/.

5. J. Morokuma et al., 'Planarian regeneration in space: persistent anatomical, behavioral and bacteriological changes induced by space travel', *Regeneration*, vol. 4, 85–102 (2017); https://doi.org/10.1002/reg2.79.

6. Michael Levin, 'The wisdom of the body: future techniques and approaches to morphogenetic fields in regenerative medicine, developmental biology and cancer', *Regenerative Medicine*, 6 (6), 667–73 (2011).

7. Ibid.

8. Recent experiments have demonstrated that a whole range of genes may be activated according to the *shape* of the cells' surroundings. This was demonstrated by incarcerating human stem cells in tiny cylinders, cubes, triangles, and so on. See Min Bao et al., '3D microniches reveal the importance of cell size and shape', *Nature Communications*, vol. 18, 1962 (2017).

9. S. Sarker et al., 'Discovery of spaceflight-regulated virulence mechanisms in salmonella and other microbial pathogens', *Gravitational and Space Biology*, 23 (2), 75–8 (August 2010).

10. J. Barrila et al., 'Spaceflight modulates gene expression in astronauts', *Microgravity*, vol. 2, 16039 (2016).

11. Lynn Caporale, 'Chance favors the prepared genome', *Annals of the New York Academy of Sciences*, 870, 1–21 (18 May 1999).

12. Andreas Wagner, *Arrival of the Fittest* (Current, 2014).

13. Krishnendu Chatterjee et al., 'The time scale of evolutionary innovation', *PLoS Computational Biology* (2014); https://doi.org/10.1371/journal.pcbi.1003818.

14. Cited by Susan Rosenberg in: Emily Singer, 'Does evolution evolve under pressure?', *Quanta* (17 January 2014); https://www.wired.com/2014/01/evolution-evolves-under-pressure/.

15. T. Dobzhansky, 'The genetic basis of evolution', *Scientific American*, 182, 32–41 (1950).

16. J. Cairns, J. Overbaugh and S. Miller, 'The origin of mutants', *Nature*, 335, 142–5 (1988).

17. Ibid.

18. Barbara Wright, 'A biochemical mechanism for nonrandom mutations and evolution', *Journal of Bacteriology*, vol. 182, no. 11, 2993–3001 (2000)

19. Susan M. Rosenberg et al., 'Stress-induced mutation via DNA breaks in Escherichia coli: a molecular mechanism with implications for evolution and medicine', *Bioessays*, vol. 34, no. 10, 885–92 (2012).

20. Jablonka, *Evolution in Four Dimensions*.

21. Caporale, 'Chance favors the prepared genome'.

22. Barbara McClintock, 'The significance of responses of the genome to challenge', The Nobel Foundation (1984); http://nobelprize.org/nobel_prizes/medicine/laureates/1983/mcclintock-lecture.pdf.

23. See, for example, Jürgen Brosius, 'The contribution of RNAs and retroposition to evolutionary novelties', *Genetica*, vol. 118, 99 (2003).

24. Wagner, *Arrival of the Fittest*, p. 5.

25. Ibid.

26. Kevin Laland, 'Evolution evolves', *New Scientist*, 42–5 (24 September 2016).

27. Deborah Charlesworth, Nicholas H. Barton and Brian Charlesworth, 'The sources of adaptive variation', *Proceedings of the Royal Society B*, vol. 284: 20162864 (2017).

28. D. Hanahan and R. A. Weinberg, 'The hallmarks of cancer', *Cell*, 100 (1): 57–70 (January 2000); doi: 10.1016/S0092-8674(00)81683-9, PMID 10647931; and 'Hallmarks of cancer: the next generation', *Cell*, 144 (5), 646–74 (4 March 2011); doi: 10.1016/j.cell.2011.02.013.

29. C. Athena Aktipis et al., 'Cancer across the tree of life: cooperation and cheating in multicellularity', *Philosophical Transactions of The Royal Society B*, 370: 20140219 (2015); http://dx.doi.org/10.1098/rstb.2014.0219. Of course, not all species are equally susceptible.

30. Tomislav Domazet-Lošo et al., 'Naturally occurring tumours in the basal metazoan Hydra', *Nature Communications*, vol. 5, article number: 4222 (2014); doi: 10.1038/ncomms5222.

31. Paul C. W. Davies and Charles H. Lineweaver, 'Cancer tumours as Metazoa 1.0: tapping genes of ancient ancestors', *Physical Biology*, vol. 8, 015001–8 (2011).

32. Tomislav Domazet-Lošo and Diethard Tautz, 'Phylostratigraphic tracking of cancer genes suggests a link to the emergence of multi-cellularity in metazoa', *BMC Biology*, 20108:66 (2010); https://doi.org/10.1186/1741-7007-8-66.

33. Anna S. Trigos et al., 'Altered interactions between unicellular and multicellular genes drive hallmarks of transformation in a diverse range of solid tumors', *Proceedings of the National Academy of Sciences*, vol. 114, no. 24, 6406–6411 (2017); doi: 10.1073/pnas.1617743114; see also Kimberly J. Bussey et al., 'Ancestral gene regulatory networks drive cancer', *Proceedings of the National Academy of Sciences*, vol. 114 (24), 6160–62 (2017).

34. Luis Cisneros et al., 'Ancient genes establish stress-induced mutation as a hallmark of cancer', *PLoS ONE*; https://doi.org/10.1371/journal.pone.0176258 (2017).

35. Amy Wu et al., 'Ancient hot and cold genes and chemotherapy resistance emergence', *Proceedings of the National Academy of Sciences*, vol. 112, no. 33, 10467–72 (2015).

36. George Johnson, 'A tumor, the embryo's evil twin', *The New York Times*, 17 March 2014.

5. SPOOKY LIFE AND
QUANTUM DEMONS

1. Richard Feynman, 'Simulating physics with computers', *International Journal of Theoretical Physics*, vol. 21, 467–88 (1982).

2. Harry B. Gray and Jay R. Winkler, 'Electron flow through metallo-proteins', *Biochimica et Biophysica Acta*, 1797, 1563–72 (2010).

3. Gábor Vattay et al., 'Quantum criticality at the origin of life', *Journal of Physics: Conference Series*, 626, 012023 (2015).

4. Ibid.

5. Gregory S. Engel et al., 'Evidence for wavelike energy transfer through quantum coherence in photosynthetic systems', *Nature*, vol. 446, 782–6 (12 April 2007); doi: 10.1038/nature05678.

6. Patrick Rebentrost et al., 'Environment-assisted quantum transport', *New Journal of Physics*, 11 (3):033003 (2009).

7. Roswitha Wiltschko and Wolfgang Wiltschko, 'Sensing magnetic directions in birds: radical pair processes involving cryptochrome', *Biosensors*, 4, 221–42 (2014).

8. Ibid.

9. Thorsten Ritz et al., 'Resonance effects indicate a radical-pair mechanism for avian magnetic compass', *Nature*, 429, 177–80 (13 May 2004); doi: 10.1038/nature02534.

10. See Mark Anderson, 'Study bolsters quantum vibration scent theory', *Scientific American*, 28 January 2013; see also https://www.ted.com/talks/luca_turin_on_the_science_of_scent and 'Smells, spanners, and switches' by Luca Turin, inference-review.com, vol. 2, no. 2 (2016).

11. An excellent critical review of the subject is given by Ross D. Hoehn et al., 'Status of the vibrational theory of olfaction', *Frontiers in Physics*, 19 March 2018; doi: 10.3389/fphy.2018.00025.

12. Scott Aaronson, 'Are quantum states exponentially long vectors?' *Proceedings of the Oberwolfach Meeting on Complexity Theory*, arXiv: quant-ph/0507242v1, accessed 8 March 2010.

13. The issue here is that there are different types of noise and their disruptive effects on quantum systems can be very different. Recent calculations have identified a regime in which a certain form of noise paradoxically *enhances* quantum transport effects. See, for example, S. F. Huelga and M. B. Plenio, 'Vibrations, quanta and biology', *Contemporary Physics*, vol. 54, no. 4, 181–207 (2013); http://dx.doi.org/10.108 0/00405000.2013.829687. See also reference 6.

14. Apoorva Patel, 'Quantum algorithms and the genetic code', *Pramana*, vol. 56, 365 (2001).

15. Philip Ball, 'Is photosynthesis quantum-ish?', *Physics World*, April 2018.

16. Adriana Marais et al., 'The future of quantum biology', *Royal Society Interface Focus*, vol. 15, 20180640 (2018).

6. ALMOST A MIRACLE

1. George Whitesides, 'The improbability of life', in: John D. Barrow et al., *Fitness of the Cosmos for Life: Biochemistry and Fine-tuning* (Cambridge University Press, 2004), p. xiii.

2. Stanley Miller, 'A production of amino acids under possible primitive Earth conditions', *Science*, vol. 117 (3046): 528–9 (1953).

3. Francis Crick, *Life Itself: Its Origin and Nature* (Simon and Schuster, 1981), p. 133.

4. Charles Darwin, letter he posted to his friend Joseph Dalton Hooker on 29 March 1863. For a contextual historical essay, see Juli Peretó, Jeffrey L. Bada and Antonio Lazcano, 'Charles Darwin and the origin of life', in: *Origins of Life and Evolution of Biospheres*, vol. 39, 395–406 (2009).

5. J. William Schopf et al., 'SIMS analyses of the oldest known assemblage of microfossils document their taxon-correlated isotope compositions', *Proceedings of the National Academy of Sciences*, vol. 115, no. 1, 53–8; doi: 10.1073/pnas.1718063115.

6. Manfred Eigen and Peter Schuster, *The Hypercycle: A Principle of Natural Self-Organization* (Springer, 1979).

7. Eugene V. Koonin and Artem S. Novozhilov, 'Origin and evolution of the genetic code: the universal enigma', *IUBMB Life*, February 2009; 61(2): 99–111 (February 2009); doi: 10.1002/iub.146.

8. Jacques Monod, *Chance and Necessity*, trans. A Wainhouse: (Alfred A. Knopf, 1971), p. 171.

9. George Simpson, 'On the nonprevalence of humanoids', *Science*, vol. 143, issue 3608, 769–75 (21 February 1964).

10. Christian de Duve, *Vital Dust: The Origin and Evolution of Life on Earth* (Basic Books, 1995).

11. Mary Voytek, quoted in Bob Holmes, 'The world in 2076: We still haven't found alien life', *New Scientist*, 16 November 2016.

12. Carl Sagan, 'The abundance of life-bearing planets', *Bioastronomy News*, vol. 7, 1–4 (1995).

13. Erwin Schrödinger, *What is Life?* (Cambridge University Press, 1944), p. 80.

14. Harold Morowitz and Eric Smith, *The Origin and Nature of Life on Earth: The Emergence of the Fourth Geosphere* (Cambridge University Press, 2016).

15. Paul C. W. Davies et al., 'Signatures of a shadow biosphere', *Astrobiology*, vol. 9, no. 2, 241–51 (2009).

16. John Maynard-Smith and Eörs Szathmáry, *The Major Transitions of Evolution* (Oxford University Press, 1995).

7. THE GHOST IN THE MACHINE

1. Werner Loewenstein, *Physics in Mind* (Basic Books, 2013), p. 21.

2. Erwin Schrödinger, *Mind and Matter* (Cambridge University Press, 1958).

3. Gilbert Ryle, *The Concept of Mind* (Hutchinson, 1949).

4. Alan Turing, 'Computing machines and intelligence', *Mind*, vol. 49, 433–60 (1950).

5. Roger Penrose, *The Emperor's New Mind: Concerning Computers, Minds and the Laws of Physics* (Oxford University Press, 1989).

6. Fred Hoyle, *The Intelligent Universe* (Michael Joseph, 1983).

7. Albert Einstein, letter to widow of his friend Michel Besso, dated 21 March 1955.

8. Lewis Carroll, *Alice's Adventures in Wonderland* (1865).

9. There are other cells too, called glial cells. They actually outnumber neurons. Their role is still not completely clear.

10. See, for example, 'Ion channels as Maxwell demons', in: Werner Loewenstein, *The Touchstone of Life* (Oxford University Press, 1999).

11. Giulio Tononi et al., 'Integrated information theory: from consciousness to its physical substrate', *Perspectives*, vol. 17, 450 (2016).

12. Larissa Albantakis, 'A tale of two animats: what does it take to have goals?', FQXi prize essay (2016).

13. Masafumi Oizumi, Larissa Albantakis and Giulio Tononi, 'From the phenomenology to the mechanisms of consciousness: integrated information theory 3.0', *PLoS Computational Biology*, vol. 10, issue 5, e1003588 (May 2014).

14. Titus Lucretius Carus, *De rerum natura*, Book II, line 216.

15. Erik Hoel, 'Agent above, atom below: how agents causally emerge from their underlying microphysics', FQXi prize essay (2017). Published in Anthony Aguirre, Brendan Foster and Zeeya Merali (eds.), *Wandering towards a Goal: How Can Mindless Mathematical Laws Give Rise to Aims and Intention?* (Springer, 2018), pp. 63–76.

16. Penrose, *The Emperor's New Mind*.

17. Ibid.

18. George F. R. Ellis, Denis Noble and Timothy O'Connor, 'Top-down causation: an integrating theme within and across the sciences?', *Royal Society Interface Focus*, vol. 2, 1–3 (2012).

19. A proposal somewhat along these lines has also been made by a group of physicists at Oxford University. See Kobi Kremnizer and André Ranchin, 'Integrated information-induced quantum collapse', *Foundations of Physics*, vol. 45, issue 8, 889–99 (2015).

20. David Chalmers, *The Character of Consciousness* (Oxford University Press, 2010).

EPILOGUE

1. Albert Einstein, letter to Leo Szilárd. See R. W. Clark, *Einstein: The Life and Times* (Avon, 1972).

2. René Descartes, *Philosophical Essays and Correspondence*, ed. Roger Ariew (Hackett, 2000), pp. 28–9.

3. Charles Darwin, *On the Origin of Species* (John Murray; 1st edn, 1859), p. 490.

4. Nigel Goldenfeld and Carl Woese, 'Life is physics: evolution as a collective phenomenon far from equilibrium', *Annual Reviews of Condensed Matter Physics*, vol. 2, 375–99 (2011).

5. Ibid.

6. Other commentators have suggested something similar. For example, Philip Ball writes, 'If there's a kind of physics behind biological teleology and agency, it has something to do with the same concept that seems to have become installed at the heart of fundamental physics itself: information.' See 'How life (and death) spring from disorder', *Quanta*, 26 January 2017, p. 44.

7. George F. R. Ellis, Denis Noble and Timothy O'Connor, 'Top-down causation: an integrating theme within and across the sciences?', *Royal Society Interface Focus*, vol. 2, 1–3 (2012).

8. Sidney Fox, 'Pre-biotic roots of informed protein synthesis', in *Cosmic Beginnings and Human Ends*, eds. Clifford Matthews and Roy Abraham Varghese (Open Court, 1993), p. 91.

9. Gary Steinman and Marian Cole, 'Synthesis of biologically pertinent peptides under possible primordial conditions', *Proceedings of the National Academy of Sciences*, vol. 58, 735 (1976).

10. A start along these lines has been made in the following paper: Ivo Grosse et al., 'Average mutual information of coding and noncoding DNA', *Pacific Symposium on Biocomputing*, vol. 5, 611–20 (2000), where they write, 'Here, we investigate if there exist species-independent statistical patterns that are different in coding and noncoding DNA. We introduce an information-theoretic quantity, the average mutual information (AMI), and we find that the probability distribution functions of the AMI are significantly different in coding and noncoding DNA.'

11. There is the even thornier question of where any laws ultimately come from, a topic that I tackled in *The Goldilocks Enigma* (2006).

12. Jacques Monod, *Chance and Necessity*, trans. A. Wainhouse (Vintage, 1971), p. 180.

13. Christian de Duve, *Vital Dust: The Origin and Evolution of Life on Earth* (Basic Books, 1995), p. 300.

Illustration Credits

1. MPI/Stringer/Getty Images
2. Prof. Arthur Winfree/Science Photo Library/Getty Images
4. Adapted from Fig. 1 of Koji Maruyama, Franco Nori and Vlatko Vedral, 'The phyiscs of Maxwell's demon and information', arXiv: 0707.3400v2 (2008)
10. Courtesy of Alyssa Adams
12. Courtesy of Michael Levin
13. Adapted from C. Athena Aktipis et al., 'Cancer across the tree of life: cooperation and cheating in multicellularity', *Philosophical Transactions of The Royal Society B*, 370: 20140219 (2015), by kind permission of the authors

Index

Page references in *italic* indicate figures and tables or their captions.

Aaronson, Scott (epigraph) 162
Adami, Christoph (epigraph) 56
Adams, Alyssa 80–82, 213
Adams, Dany 115
adenosine diphosphate (ADP) 57–8
adenosine triphosphate (ATP) 57–
    8, 59, 61, 150
African clawed frog (*Xenopus*)
    114–15
agency, and human consciousness
    201–5
Aktipis, Athena 132
Albantakis, Larissa 200
alpha decay 149
amino acids 17, 18, 19, 20, 21, 38,
    64, 124, 149
    arginine 211
    Fox's claims 214
    and the origins of life 166–7,
        174, 176
    peptides 214
aminoacyl-tRNA synthetase 19
analytical (Jungian) psychology 187
anasthesia 206
animal rights 185
anti-particles 165
antlers, deer 119–20
ants 25, 83, 101–4
aperiodic crystals 25, 26

Archean chert 168
Aristotle 8
arrival of the fittest 123–6
artificial intelligence (AI) 86
    and consciousness 185, 186–7
asteroids 10–11, 169–70, 181
astrobiology 13–14, 176–7, 179
Atacama Desert 11–12
'atomic swerve' 201–2
ATP *see* adenosine triphosphate
Austin, Robert 142
autocatalysis 173–4
axons 196, 197

bacteria 1, 15, 65, 87, 91, *139*
    adaptive mutation response to
        stress 141
    cell cycles 94
    Desulforudis audaxviator 11
    E. coli 39n, 62, 88, 123–4
    green sulphur 154
    mycoplasma genitalium 178
    reproduction and heat 62
    salmonella 119
Barzel, Baruch 100–101
bees 25, 83
Bennett, Charles 48–9, 54
    analysis of Maxwell's demon 48–9
bio-engineering 91–3

bioinformatics 35, 110
biology
  astrobiology 13–14, 176–7, 179
  biological circuits 87–93
  biological information 7–8, 67,
    74, 99, 109, 175, 183; see also
    genetic code
  complexity 214
  information as bridge between
    physics and 1–2, 186, 209–17
  molecular 6, 10; see also organic
    molecules
  and physical laws 211–12, 213
  quantum see quantum biology
  reductionism 6, 83–5
  synthetic biology 91–3, 179
  systems biology 91–2
  and Turing 71–2, 73
  see also life
biomachines 56–64
birds 25
  Arctic tern 157
  blackpoll warbler 157
  eyes 158–9
  navigation mechanisms 156–70
  quantum biology and bird
    navigation 157–60
black boxing 199–200, 203
BLADE (Boolean logic and
    arithmetic through DNA
    excision) 91
Bletchley Park 70, 71
Bohr, Niels 6, 162
Boole, George 96n
Boolean networks 96n
Bornholdt, Stefan 94
brain
  compartmentalization 203
  connections and complexity 196;
    see also neurons

  and consciousness 187–9, 190–
    91, 192, 195–8, 205–8
  Earth's magnetic field and birds'
    brains 158–9
  flow of electricity 196–8
  free will, agency, and the brain
    201–5
  gated ion channels 195–8
  as information-processor 25, 183,
    195–6
  neurons see neurons
  neuroscience 187, 190,
    194–5, 201
  and quantum mechanics 202,
    205–8
  synapses see synapses
Bussey, Kimberly 140, 141

Cairns, John 123–4
cancer 91, 114, 118, 119
  as an atavistic phenotype 138–43
  COSMIC inventory of cancer
    genes 140, 141
  DAVID database of cancer
    genes 141
  doxorubicin resistance 142
  evolutionary roots 136–43,
    138, 139
  and genomic transpositions 128
  hallmarks 131, 141, 142
  metastasis 131, 136
  mutations 131, 140
  and reactive oxygenic molecules
    170–71
  relation to multicellularity 132–6,
    138, 140, 141–2
  across tree of life 132, 133, 139
  tumour suppressor genes 135
  tumours 114, 119, 135, 138–9,
    141, 142

Caporale, Lynn 126
  epigraph 121
carbon 16, 171
  C^{12} isotope 169
  carbon dioxide 11, 16, 20, 151
Carroll, Lewis: *Alice's Adventures in Wonderland* 195
Cartesian dualism 185–6, 191
Cassini spacecraft 13, 15
causation
  causal power of information 35, 47
  top-down 74, 215
cell cycle 94–101, 95, 96
cellular automata (CA) 75–82, 77, 81, 82, 213
  see also Game of Life
Chaitin, Gregory (epigraph) 67
Chalmers, David 208
chaotic systems 214
chemical complexity 15, 17, 67
chemical energy 9, 28, 58, 63
chemical networks 90, 106, 167
chemical pathways 13, 87, 90, 95, 111–12, 172
chromatin 90–91
chromosome reorganization 127–9
chromothripsis 128
Chroococcidiopsis 12
Cisneros, Luis 140, 141
coherence, quantum 147, 155, 156, 163–4
Cole, Marian 214
comets 10–11, 169–70
complexity
  of biology/life 7–8, 14, 72, 75, 82, 166, 196, 214
  of the brain 196
  chemical 15, 17, 67
  consciousness and pattern complexity 191

of dissipative structures 22
evolution of 80, 82
and Game of Life 76–80
informational 7–8
and open-ended evolvability 79–82
and state-dependent dynamics 80–82, 213–17
computation 45–6, 66, 67–72
  bioinformatics 35, 110
  by cells 64–6, 89
  distributed 103
  quantum 144–5, 148, 164–5
  reversible 48–9
  and self-reference 69–72, 73, 212–13
  and Turing 70–72, 73
  universal 70–75, 77, 212
computers 35–6, 86–7, 109–10
  Colossus 71
  and Game of Life 75–80, 77
  heat generated by 45–6, 48
  quantum 144–5, 148, 164
  and the simulation argument 189
  software 87, 93, 109–10
  Turing and a universal computer 70–71, 73, 77
  universal constructors (von Neumann machines) 72–5
consciousness/mind 183, 184–208
  and agency 201–5
  and AI 185, 186–7
  and animal pain/rights 185
  and the brain 187–9, 190–91, 192, 195–8, 205–8
  and Cartesian dualism 185–6
  outside the body 188–9
  and the flow of time 192–5
  and free will 201–5

consciousness/mind – *cont.*
  and information 198–201, 206–8
  measuring consciousness
    199–201
  mind–body problem 187, 191
  and mind over matter 191, 205–6
  popular views of 187
  and quantum mechanics 202,
    205–8
  self-consciousness 191
  simulation argument 189
contact inhibition 118–19
convection cells 22
Conway, John, Game of Life 75–80,
  77, 93
cosmic imperative, life as a 176,
  179, 181, 215n, 217
Crick, Francis 25, 167
CRISPR technology 178
Cronin, Lee 17n
cross-breeding 178
cryptochromes 159

Darwin, Charles 1, 7, 72
  On the Origin of Species
    167–8, 211
Darwinism 6, 61, 110, 120, 122,
  167–8, 180
  'Darwinism 2.0' 110–43
  heritable errors as drivers of
    Darwinian evolution 61, 75, 110
  intersection with information
    theory 108
  'molecular Darwinism' 173n
Davidich, Maria 94
Davidson, Eric 106–7
Dawkins, Richard 83
  'Mount Improbable' 14–15, 173
de Duve, Christian 176, 181,
  215n, 217

de Valera, Éamon 5
decoherence, quantum 147, 206
deer antlers 119–20
demons
  applied demonology 54–6
  and Bennett 48–9
  biological quantum demons 148,
    155, 160–62, 163
  in the brain's wiring 195–8, 204
  and chromatin 91
  Descartes' demon 188
  in the genes 126–30
  and information engines 49–54,
    50, 51
  and Landauer 45, 48
  in living cells 56–66, 91, 126–30
  Maxwell's demon 27–35, 34,
    54–6, 148, 164, 193
  and nanotechnology 54–6
  and Szilárd 42–4, 50
dendrites 196
Descartes, René 185, 187, 188, 210
Desulforudis audaxviator 11
determinism 202
  over-determinism 202, 203
deuterium 161–2
Deutsch, David 144
  epigraph 10
dissipative structures 22, 23
DNA 6, 17–20, 23–4, 61, 74
  as an aperiodic crystal 25
  chromatin 90–91
  damage from oxyenic
    molecules 171
  'dark sector' 113
  double-strand breaks 124, 141
  errors in replication 25, 61, 75,
    110, 127
  genetic code *see* genetic code
  genomic transpositions 127–9

genotypes 105
and information theory 2, 25, 174
informational content of a
   genome 38-9, 40
junk 65, 113
and 'natural genetic
   engineering' 130
nucleosome spacing 112-13
phenotypes *see* phenotypes
polymerase 61-2
promoters 88
and quantum computation 164
and quantum tunnelling 149-51
repair 124-5, 127, 129, 141
replication 25, 61-2, 74, 75, 94,
   110, 127
sequencing 149, 171-2
structure of 82-3, 104, 166-7
reverse transcription 129
and transcription factors 87-9
Dobzhansky, Theodosius 122-3
epigraph 109
dogs 178
dopamine 198
doxorubicin 142
Dreisch, Hans 8-9, 105, 107
dualism
   Cartesian 185-6, 191
   and computational theory 186
   and the 'ghost in a machine' 191
dynein 60-61

Earth
   accretion 10-11, 170
   'anti-accretion' 11, 181
   DNA and the story of life on
      17-20; *see also* DNA
   magnetic field, and bird
      navigation 157-60
   shadow biosphere 181-3

ecosystems 25
Eddington, Sir Arthur 32n
effective information theory 203-4
EGF (epidermal growth factor)
      molecules 83-4
Einstein, Albert 145-6, 147
   epigraph 209
electricity/electrical energy 10,
      63, 113
   and the brain 196-8
   electric cell polarization
      114-15, 197
   electro-transduction 112-18
   and quantum tunnelling 149-51
electron tunnelling 146, 149-51,
      161-2
embryo development 8-9, 64, 105,
      107, 111
   and cancer 142
   and electro-transduction 114-18
   and mechano-transduction
      118-19
   and 'morphogenetic fields' 105,
      114n
   and mystery of morphogenesis
      104-7
emergence 83, 217
Enceladus 15-16
energy
   ATP energy molecule 57-8, 59,
      61, 150
   chemical 9, 28, 58, 63
   comparisons with information 47
   conservation of 29n
   converted into other forms by
      nano-machines 63
   and dissipative structures 22
   electrical *see* electricity/electrical
      energy
   ethereal 8-9

energy – *cont.*
  flow 26
  heat *see* heat
  and industrial growth 28
  interplay of information, entropy
    and 209
  interplay with information in
    living organisms 56–63
  kinetic 30, 47
  and life 2, 7, 8–9, 10, 12, 13, 20,
    23, 26
  mechanical 9, 28, 63
  and Newton's laws 28
  from nuclear radiation 11
  and photosynthesis 12, 154–6
  and resonance 160
  storage 26, 155
  and Szilárd's engine 42–4, 43, 55,
    68–9
Engel, Greg 151, 154–6
England, Jeremy 62
entanglement (spooky action-at-a-
    distance) 146
  and bird navigation 159–60
  refrigeration and entangled
    quantum particles 56n
entelechy 9
entropy 30–32, 39–40
  and the arrow of time 30, 31, 32,
    192–3
  and disorder 5–6, 31, 32, 39, 148
  from erasure acts 46, 49, 53–4, 55
  and information engines 49–54,
    50, 55
  interplay of information, energy
    and 209
  and Landauer limit 46, 48–9
  and quantum effects 148
  and ratchets 60
  and Szilárd's engine 42–4, *43*

transfer entropy 98–9
  see also second law of
    thermodynamics
epidermal growth factor (EGF)
    molecules 83–4
epigenetics 83, 110, 111–21, 126,
    129, 130, 142
  epigenetic control 74
  epigenetic inheritance 116,
    120–21
Escherichia coli 39n, 62, 88, 123–4
eukaryotic cells 90–91, 94–7, *95*, *96*
evolution 61, 109, 211
  and the arrival of the fittest
    123–6
  cancer's deep evolutionary roots
    136–43, *138*, *139*
  of complexity 80, 82
  complexity levels necessary for
    open-ended evolvability 79–82
  Darwinian theory of 6, 61, 75,
    110, 120, 122, 167–8, 180; *see
    also* Darwinism
  of evolvability 125–6
  modern synthesis 6
  and genomic transpositions
    127–9
  heritable errors as drivers of
    Darwinian evolution 61, 75, 110
  of information-processing rules
    82, 213–17
  Laland's 'Extended Evolutionary
    Synthesis' 130
  Lamarckian theory of 120, 121–
    2; *see also* Lamarckism
  and mutation *see* mutation
  and 'natural genetic
    engineering' 130
  and natural selection *see* natural
    selection

oxygen and evolution of aerobic life 170
replication processes' evolvability 74–5
and self-reference 212–13
excitons 155, 156
Eyemouth prawns 185n

Faraday, Michael 105, 158
feed-forward 85, 87, 88, 89
loops 90
feedback 56, 85, 87, 88, 200
loops 64, 106, 174, 200
negative 103
Feynman, Richard 39n, 54n, 100, 144
Feynman's ratchet 59, 60
fission yeast 94–7, 95, 96
flatworms 115–17, 116
two-headed 111, 116, 116, 119, 122
Fleming, Graham 154–6
Flowers, Tommy 71
FMO complex 154–5
fossils 168–9
Fox, Sidney 214
free will 201–5
fruit fly (Drosophila) 105–6, 162

Galileo Galilei 210
Game of Life 75–80, 77, 93
Gamow, George 21
Gard, Charlie 184n
gene networks 93–101, 95, 96, 106–7
genetic code 6, 18–20, 21, 64, 174–5, 211
biological information as software of life 7–8, 67, 74, 99, 109, 175, 183

contextuality of genetic information 65
a genome's storage of information 38–9, 104
and measurement of information 38–40
and mutation 35; see also mutation
genetics
and biological reductionism 83
and CRISPR technology 178
genetic sequencing 18–19, 20, 25, 38, 40, 61, 65, 87–8, 89, 110, 116–17, 129, 139–40, 149, 171–2; see also genetic code
GM foods 178
and information see genetic code; information
Mendel's laws of 211
modern synthesis 6
'natural genetic engineering' 130
synthetic circuits 91–2
genotypes 105
glycolaldehyde 17
GM foods 178
Gödel, Kurt, theorem 69–70, 73
Gold, Thomas 171n
Goldenfeld, Nigel 213
Goode, David 139–40
Gray, Harry 150
Greene, Dave 79
Greenland 169
growth factors 106
epidermal 83–4

halting problem 68, 71
Hameroff, Stuart 206
Hartwell, Lee 94
Harush, Uzi 100–101

heat 2, 8, 9, 27, 39
    converted into work 49–54,
        50, 51
    engines 7, 30
    and entropy *see* entropy
    from erasure acts 46
    extraction in refrigeration
        30–31, 56
    flow 29, 30
    generated by computers 45–6, 48
    harnessing of 28–9
    and information engines 49–54,
        50, 51
    Maxwell and the theory of 27
    Maxwell's demon and 28–34, *34*
    and microbes 12, 16
    from reproduction of bacteria 62
    and Szilárd's engine 42–4, *43*
    thermodynamics *see*
        thermodynamics
Heisenberg, Werner 6
Helmholtz, Hermann von 9
heptanol 115–16
Higgs boson 165
Hilbert, David 67–9
hinnies 120
histones 111, 112
Hoel, Erik 203–4
Hoffmann, Peter (epigraph) 27
Hoyle, Fred 188–9
Huygens, Christiaan 13
hydrogen 11, 161, 162, 171
hyperthermophiles 171–2

IBM 45, 46
immune system 123, 131 and n, 135
immunosuppression 135
immunotherapy 131n
indeterminism, quantum 145,
    202, 205

Industrial Revolution 28, 30
infinity 41n, 205
information
    and agency/free will 202–5
    reverse transcription 129
    bioinformatics 35, 110
    biological information as
        software of life 7–8, 67, 74, 99,
        109, 175, 183
    bits/bytes 36, 37–9, 40, 66, 92,
        145, 198; *see also* qubits
    brain as information-processor
        25, 183, 195–6
    as bridge between physics and
        biology 1–2, 186, 209–17
    causal power of 35, 47
    coded in DNA *see* genetic code
    comparisons with energy 47
    complexity 7–8
    and consciousness 198–201,
        206–8
    digital 61, 92–3, 144, 174; *see
        also* computers
    effective information theory
        203–4
    engines 49–56, *50, 51*, 63
    erasure 46, 49, 53–4, *55*, 103
    evolution of information-
        processing rules 82, 213–17
    flow 2, 41, 64, 76, 78, 85, 86,
        89–90, 92, 98–101, 103–4,
        107–8, 111–12, 117, 130, 142,
        167, 190, 192, 201, 214
    genetic *see* genetic code
    heritable 108
    informational basis of life 2, 6–8,
        24–6, 65–6, 67, 74, 104, 172,
        174, 211
    integrated 199–201, 203, 206–7
    and Landauer limit 46, 48–9

management 64, 92, 144, 164–5,
    172, 174
measuring 35–41, 97n
morphological 117–18
neural 190–201
patterns 2, 46, 47, 67, 73, 76–9,
    80–81, 81, 86, 89, 92, 97–8,
    99–101, 107–8, 109, 110, 112–
    13, 118, 166, 167, 186, 190–92,
    195, 207–8, 213, 214, 215–16
and physical laws 41–2, 179,
    210–17
physical nature of 46–7
positional 41–4, 43, 55
processing 25, 44–53, 65–6, 82,
    87, 172, 183, 190, 195–6, 201,
    213–17
quantum mechanics and
    information management 144,
    164–5
qubits 144–5, 148, 164
informational architecture 183
replication of heritable
    information 24–5
semantic 167
Shannon's definition 65
storage 25, 35, 38–9, 40,
    46–7, 104, 117–18, 130,
    172, 175, 195
autonomous existence of 47, 78
and Szilárd's engine 42–4, 43, 55
technology 35–6, 44–5; see also
    computers
theory 2, 25, 65–6, 103, 108, 174,
    202–4, 207
transfer 98–9
universality in informational
    organization 214–17
integrated information 199–201,
    203, 206–7

International Space Station 111
intrinsic spin of electrons 158

Jablonka, Eva 126
    epigraph 111
Jarzynski, Christopher, information
    engines 49–54, 50, 51
Johnson, George 142
Jungian psychology 187
Jupiter, Great Red Spot 21, 22

Kim, Hyunju 98
kinesin 57–8, 60–61, 63, 64
kinetic energy 30, 47
Kobayashi Hideki 91–2
Koonin, Eugene 174–5
Küppers, Bernd-Olaf
    (epigraph) 24

Laland, Kevin 130
Lamarck, Jean-Baptiste 121
Lamarckism 120, 121–6
Landauer, Rolf 45–6, 61
    and Maxwell's demon 45, 48
Landauer limit 46, 48–9, 196
Lazebnik, Yuri 83–5
Leche, Johannes 156
Leigh, David 54
Levin, Gil 16
Levin, Michael 111, 114, 115–18
life
    biology see biology
    chemical and informational
        amalgam of 67
    chemistry's rigging in favour of
        214–15
    complexity of 7–8, 14, 72, 75, 82,
        166, 196, 214
    and the 'cosmic imperative' 176,
        179, 181, 215n, 217

life – *cont.*
demons in living cells as engines
of 56–66, 91
and dissipative structures 22, 23
DNA and the story of life on
Earth 17–20; *see also* DNA
elements essential to 171
embryo development *see* embryo
development
and energy *see* energy: and life
evolution of *see* evolution
extraterrestrial life 175–7
force 8–9, 105, 113
and Game of Life 75–80, 77
informational basis of 2, 6–8,
24–6, 65–6, 67, 74, 104, 172,
174, 211–12
informational patterns of 2, 67,
78–9, 86, 89, 97–8, 100, 107–
8, 109, 112–13, 118, 166, 167,
190, 192, 207–8, 215–16
laboratory creation of 177–9
'life meter' 13–16
first appearance on Earth 169–72
logic of 67–108
and Maxwell's demon *see*
demons: Maxwell's demon
morphogenesis *see*
morphogenesis
mutation *see* mutation
and nano-machines 56–64
organic molecules *see* organic
molecules
organisms as prediction
machines 65
origins of 166–83, 215–17
as a planetary phenomenon 180
reconceptualization of 85
and a shadow biosphere 181–3
single-celled organisms 132, 136

'software' of 7–8, 67, 74, 99, 109,
175, 183
thermodynamic efficiency
62–3, 166
tree of *139*, 189–90
limb regeneration 119
Lindsay, Stuart 149
Lizier, Joseph 78
logic 68, 69–70
axioms and deductions 68,
69–70
gates 48
of life 67–108
Lucretius 201–2

macrophages 135
magnetic field of Earth, and bird
navigation 157–60
Mars 12, 16–17, 169–70, 176
mathematics 36–8, 65, 67–72, 97,
134, 213
logical foundations of 68–9
patterns buried in chaotic
systems 214
and undecidability 69, 70, 71
Maxwell, James Clerk 27, 28
letter to Tait 27, 28, 32
Maxwell's demon 27–35, *34*, *50*,
54–6, 148, 164, 193
theory of heat 27
McClintock, Barbara 127–9, 130
mechanical energy 9, 28, 63
mechano-transduction 118–19
metabolism 23, 99, 131, 150, 169,
170, 180, 182
lactose 88, 131
metallo-proteins 150
metastasis 131, 136
meteorites 17, 170, 181
methane 13–14, 16

microbes 39n, 119, 181–2
  archaea *139*
  bacteria *see* bacteria
  Chroococcidiopsis 12
  transfer between between Earth
    and Mars 170
  Desulforudis audaxviator 11
  in Earth orbit 119
  extremophiles 12–13, 171–2
  and heat 12, 16
  hyperthermophiles 171–2
  on Mars 16–17
  psychrophiles 94n
  radiation-resilient 170
  and a shadow biosphere 181–3
  and stromatolites 168
microbiome 119
microchips 46, 86
Miescher, Friedrich 18
Miller, Stanley 166–7
mind *see* consciousness/mind
molecular biology 6, 10
  see also organic molecules
molecular ratchets 58, 60, 61
Monod, Jacques 175, 217
Morowitz, Harold 180
morphogenesis 104–7
  and electrical pre-patterning 115
  and electro-transduction 112–18
  embryo development *see* embryo
    development
  'morphogenetic fields' 105, 114n
  morphological
    information 117–18
  regeneration 115–18, 119
morphogens 105, 106, 112
Mponeng gold mine, South
    Africa 11
mules 120
multicellularity 183

cancer as price of 132–6, 138,
    140, 141–2; *see also* cancer
  evolution of 132 and n, *133*,
    139–40
mutations 35, 90
  adaptive 124–7, 141
  and arrival of the fittest 123–6
  and cancer 131, 140
  'cold-spots 141, 142
  random 122
  and genomic transpositions
    127–9
  'hotspots' 125, 141
  somatic mutation theory 135–6
Muybridge, Eadweard 39n
mycoplasma genitalium 178

nano-machines, cellular 56–64
nanotechnology 2, 7, 25n, 209
  and applied demonology 54–6
NASA 12, 16–17, 119
natural selection 14, 90, 108, 109,
    110, 122, 126, 136, 173, 216
  arrival of the fittest 123–6
  and quantum effects 148
  survival of the fittest 122, 136
Nature 54, 85
negative feedback 103
networks
  applied to regulation of gene
    expression 94–101, *95*, *96*
  theory of 93–4, 99
  and social insects 101–4
Neumann, John von *see* von
    Neumann, John
neurons 187, 190, 195–8, 202, 203,
    204, 206
  neural information 190–201
neuroscience 187, 190, 194–5, 201
neurotransmitters 159, 198

Newton, Isaac 28, 210
Nickerson, Cheryl 119
Nixon, Richard 132
Novozhilov, Artem 174–5
nuclear radiation 11
nucleic acids 23–4
    see also DNA; RNA
Nurse, Paul 85–6, 94, 107

octanol 115–16
omega minus particle 165
optical tweezers 62
organic molecules 13, 14, 15–16, 17,
    158, 160, 170, 172, 174, 176
    nucleic acids *see* DNA; nucleic
        acids; RNA
    proteins *see* proteins
    electron tunnelling 146, 149–51
orthologs (of genes) 141n
oxygen 10, 11, 16, 20, 170, 171
    and evolution of aerobic life 170

pain, animal 185
panpsychism 187, 198
paradoxes of self-reference 7
Pauli, Wolfgang 122
Penrose, Roger 206
peptides 214
perpetual motion 32, 43–4, 48–9,
    68–9
phenotypes 105, 117
    cancer as an atavistic phenotype
        138–43
phosphorus 171
photoelectric effect 155
photons 146, 147, 151, 154, 155,
    158–9
photosynthesis 12
    and quantum biology 151–6
phylostratigraphy 139

physical laws, concepts of 41–2,
    179, 210–17
pigeons 157
Pilbara rocks 168–9
planaria 115–17
plutonium 11
Pmar1 107
potassium ions 197, 204
Pratt, Stephen 102
pre-biotic synthesis
    experiments 172
precession 158
Prigogine, Ilya 22, 23
primates 25
probability 31, 36–8, 40, 59,
    175, 177
    Dawkins' 'Mount Improbable'
        14–15, 173
    and quantum mechanics 202
Prokopenko, Mikhail 78
proteins 18–19, 20, 22, 23–4,
    56, 63, 65, 74, 106, 129,
    130, 214
    cryptochromes 159
    histones 111, 112
    metallo-proteins 150
    production of 18–19, 20, 38, 64,
        65, 74, 83–4, 87–8; *see also*
        amino acids
    in gated ion channels 196,
        197, 204
    transcription factors 87–9
psychrophiles 94n

qualia 208
quantum biology 146–65
    and bird navigation 156–60
    demons 148, 155, 160–62, 163
    and entanglement 146, 159–60
    and photosynthesis 151–6

and sense of smell 160–62
and tunnelling 146, 149–51,
161–2
quantum computers 144–5, 148, 164
quantum mechanics 25, 144–50,
162–5, 215
and biology *see* quantum biology
and the brain 202, 205–8
coherence 147, 155, 156, 163–4
collapse of the wave function
205, 207, 211
and consciousness/mind 202,
205–8
decoherence 147, 206
entanglement *see* entanglement
(spooky action-at-a-distance)
and free will 202
fuzziness 153, 205–6
indeterminism 145, 202, 205
and information management
144, 164–5
and integrated information
206–7
interference 153–4
and measurement 210–11
multiple paths 151–3, *152*
multiplicity of quantum worlds 153
Schrödinger's equation 5, 147,
149, 205, 207, 211
superposition 144–5
tunnelling 146, 149–51, 161–2
wave-particle duality 147, 151–2
weirdness of 145–8, 202, 205
qubits 144–5, 148, 164

radioactivity 11
alpha decay 149
ratchets
and entropy 60
Feynman's ratchet 59, *60*

molecular ratcheting 57–61
rectification 58
reductionism 6, 83–5, 204–5, 216
refrigerators 30–31, 56
regeneration 115–18
limb 119
reproduction/replication 7, 8, 9, 24,
71–2
complexity levels necessary for
replication and open-ended
evolvability 79–82
DNA 25, 61–2, 74, 75, 94, 110, 127
evolvability of replication
processes 74–5
Game of Life's self-reproduction 79
sexual 127
and heat in bacteria 62
of heritable information 24–5
and self-reference 73, 212–13
and universal constructors 72–5
resonance 160
reverse transcription 129
ribosomes 19, 38, 64, 74, 174
Ritz, Thorsten 159–60
RNA 23–4, 61, 62n, 113, 130, 175
BCi RNA 129
error correction 61
micro-RNA 111
mRNA 18–19, 38, 90, 107, 112, 113
polymerase 61, 87–8
reverse transcription 129
RNAi 112
and transcription factors 87–9
tRNA 19, 64, 174
RNA world theory 173n, 175
robins 159
robots 73, 104, 185, 200, 203
Rosenberg, Susan 124–5
Russell, Bertrand 69
Ryle, Gilbert 185

Sagan, Carl 177
salamanders 119
salmonella bacteria 119
scaling laws 104
Schizosaccharomyces pombe 94–7, 95, 96
Schrödinger, Erwin 6, 26, 108, 147, 149, 162–3, 179, 184, 191
    What is Life? 1, 5, 9–10, 24–5, 209
    wave equation 5, 147, 149, 205, 207, 211
sea urchins 9, 107
self-consciousness/self-awareness 191
self-reference 7, 69–72, 73, 212–13
    state-dependent dynamics 80–82, 213–17
serotonin 198
SETI 175
Shannon, Claude 36–9, 65–6, 97n, 174
Shapiro, James 129–30
'signalling' molecules 25
Simpson, George 175
simulation argument 189
smell, sense of 160–62
Smith, Eric 26, 180
social insects 101–4
    see also ants; bees
social structures 25
sodium ions 196–7, 204
solar system 11, 169–70
somatic mutation theory 135–6
space worms 111, 119
state-dependent dynamics 80–82, 213–17
Steinman, Gary 214
stigmergy 102n
stress 118

adaptive response to 126, 137, 141
    and cancer 137, 141
stromatolites 168
sulphur 171
    green sulphur bacteria 154
superposition 144–5
survival of the fittest 122, 136
synapses 197
    synaptic cleft 197–8
synthetic biology 91–3, 179
systems biology 91–2
Szilárd, Leo 41–4
    and the demon 42–4, *50*
Szilárd's engine 42–4, *43*, 55, 68–9

Tait, Peter Guthrie 27, 28, 32
Tay-Sachs syndrome 83
teleology 8, 173
telepathy 188
Tennyson, Alfred, Lord 109
theology 210, 216
thermodynamics 7, 29, 39
    and the arrow of time 30, 31, 32, 192–3
    efficiency of biological processes 62–3, 166
    first law of 29n
    heat engines 7, 29–33,
    and the laws of computing 45–6
    second law of 6, 29–32, 44, 48–9, 58, 64, 148
    refrigerators 30–31, 56 and n
    see also entropy
thought experiments 33, 41–4, 50, 60, 68–9
time
    arrow of 192–3
    the flow of 192–5

and second law of
    thermodynamics 30, 31
Titan 13–14
Tononi, Giulio 199–200, 203,
    206–7
transcription factors 87–9
transfer entropy 98–9
Trigos, Anna 139–40
tumour suppressor genes 135
tumours 114, 119, 135, 138–9,
    141, 142
tunnelling, quantum 146, 149–51,
    161–2
Turin, Luca 161–2
Turing, Alan 70–72, 73,
    106, 186
Turing machine 70–71, 73, 77
Turing test 186–7

uncertainty 36, 38, 40–41, 44n
    quantum indeterminism 145,
    202, 205
undecidability 69, 71, 72, 77–8
universal computer see Turing
    machine
universal constructors (UCs) 72–5

Vattay, Gábor 150
Venter, Craig 39n, 54n, 177, 178

Venus 170
Viking spacecraft 16–17
vitalism 8, 105
von Neumann, John
    cellular automaton 75, 78–9, 80
    and mind 205
    universal constructors 72–5
Voytek, Mary 176
Vries, Hugo de 122

Wade, Andrew J. 79
Wagner, Andreas 65, 130
Walker, Sara 2, 76n, 80–82,
    98, 213
Watson, James 25
wave-particle duality 147, 151–2
Weismann, August 120
Whitesides, George (epigraph) 166
Winkler, Jay 150
Wolfram, Stephen 80
    CA rules 80, 81, 82
World Wide Web 26
worms (planaria) 115–17
    two-headed 116–17, 116
Wright, Barbara 124, 125
Wu, Amy 142

yeast, fission, cell cycle of 94–7,
    95, 96

ALLEN LANE
an imprint of
PENGUIN BOOKS

Also Published

Ivan Krastev and Stephen Holmes, *The Light that Failed: A Reckoning*

Alexander Watson, *The Fortress: The Great Siege of Przemysl*

Thomas Penn, *The Brothers York: An English Tragedy*

David Abulafia, *The Boundless Sea: A Human History of the Oceans*

Dominic Sandbrook, *Who Dares Wins: Britain, 1979-1982*

Charles Moore, *Margaret Thatcher: The Authorized Biography, Volume Three: Herself Alone*

Orlando Figes, *The Europeans: Three Lives and the Making of a Cosmopolitan Culture*

Naomi Klein, *On Fire: The Burning Case for a Green New Deal*

Hassan Damluji, *The Responsible Globalist: What Citizens of the World Can Learn from Nationalism*

John Sellars, *Lessons in Stoicism: What Ancient Philosophers Teach Us about How to Live*

Peter Hennessy, *Winds of Change: Britain in the Early Sixties*

Brendan Simms, *Hitler: Only the World Was Enough*

Justin Marozzi, *Islamic Empires: Fifteen Cities that Define a Civilization*

Bruce Hood, *Possessed: Why We Want More Than We Need*

Frank Close, *Trinity: The Treachery and Pursuit of the Most Dangerous Spy in History*

Janet L. Nelson, *King and Emperor: A New Life of Charlemagne*

Richard M. Eaton, *India in the Persianate Age: 1000-1765*

Philip Mansel, *King of the World: The Life of Louis XIV*

James Lovelock, *Novacene: The Coming Age of Hyperintelligence*

Mark B. Smith, *The Russia Anxiety: And How History Can Resolve It*

Stella Tillyard, *George IV: King in Waiting*

Donald Sassoon, *The Anxious Triumph: A Global History of Capitalism, 1860-1914*

Elliot Ackerman, *Places and Names: On War, Revolution and Returning*

Johny Pits, *Afropean: Notes from Black Europe*

Jonathan Aldred, *Licence to be Bad: How Economics Corrupted Us*

Walt Odets, *Out of the Shadows: Reimagining Gay Men's Lives*

Jonathan Rée, *Witcraft: The Invention of Philosophy in English*

Jared Diamond, *Upheaval: How Nations Cope with Crisis and Change*

Emma Dabiri, *Don't Touch My Hair*

Srecko Horvat, *Poetry from the Future: Why a Global Liberation Movement Is Our Civilisation's Last Chance*

Paul Mason, *Clear Bright Future: A Radical Defence of the Human Being*

Remo H. Largo, *The Right Life: Human Individuality and its role in our development, health and happiness*

Joseph Stiglitz, *People, Power and Profits: Progressive Capitalism for an Age of Discontent*

David Brooks, *The Second Mountain*

Roberto Calasso, *The Unnamable Present*

Lee Smolin, *Einstein's Unfinished Revolution: The Search for What Lies Beyond the Quantum*

Clare Carlisle, *Philosopher of the Heart: The Restless Life of Søren Kierkegaard*

Nicci Gerrard, *What Dementia Teaches Us About Love*

Edward O. Wilson, *Genesis: On the Deep Origin of Societies*

John Barton, *A History of the Bible: The Book and its Faiths*

Carolyn Forché, *What You Have Heard is True: A Memoir of Witness and Resistance*

Elizabeth-Jane Burnett, *The Grassling*

Kate Brown, *Manual for Survival: A Chernobyl Guide to the Future*

Roderick Beaton, *Greece: Biography of a Modern Nation*

Matt Parker, *Humble Pi: A Comedy of Maths Errors*

Ruchir Sharma, *Democracy on the Road*

David Wallace-Wells, *The Uninhabitable Earth: A Story of the Future*

Randolph M. Nesse, *Good Reasons for Bad Feelings: Insights from the Frontier of Evolutionary Psychiatry*

Anand Giridharadas, *Winners Take All: The Elite Charade of Changing the World*

Richard Bassett, *Last Days in Old Europe: Triste '79, Vienna '85, Prague '89*

Paul Davies, *The Demon in the Machine: How Hidden Webs of Information Are Finally Solving the Mystery of Life*

Toby Green, *A Fistful of Shells: West Africa from the Rise of the Slave Trade to the Age of Revolution*

Paul Dolan, *Happy Ever After: Escaping the Myth of The Perfect Life*

Sunil Amrith, *Unruly Waters: How Mountain Rivers and Monsoons Have Shaped South Asia's History*

Christopher Harding, *Japan Story: In Search of a Nation, 1850 to the Present*

Timothy Day, *I Saw Eternity the Other Night: King's College, Cambridge, and an English Singing Style*

Richard Abels, *Aethelred the Unready: The Failed King*

Eric Kaufmann, *Whiteshift: Populism, Immigration and the Future of White Majorities*

Alan Greenspan and Adrian Wooldridge, *Capitalism in America: A History*

Philip Hensher, *The Penguin Book of the Contemporary British Short Story*

Paul Collier, *The Future of Capitalism: Facing the New Anxieties*

Andrew Roberts, *Churchill: Walking With Destiny*

Tim Flannery, *Europe: A Natural History*

T. M. Devine, *The Scottish Clearances: A History of the Dispossessed, 1600-1900*

Robert Plomin, *Blueprint: How DNA Makes Us Who We Are*

Michael Lewis, *The Fifth Risk: Undoing Democracy*

Diarmaid MacCulloch, *Thomas Cromwell: A Life*

Ramachandra Guha, *Gandhi: 1914-1948*

Slavoj Žižek, *Like a Thief in Broad Daylight: Power in the Era of Post-Humanity*

Neil MacGregor, *Living with the Gods: On Beliefs and Peoples*

Peter Biskind, *The Sky is Falling: How Vampires, Zombies, Androids and Superheroes Made America Great for Extremism*

Robert Skidelsky, *Money and Government: A Challenge to Mainstream Economics*

Helen Parr, *Our Boys: The Story of a Paratrooper*

David Gilmour, *The British in India: Three Centuries of Ambition and Experience*

Jonathan Haidt and Greg Lukianoff, *The Coddling of the American Mind: How Good Intentions and Bad Ideas are Setting up a Generation for Failure*

Ian Kershaw, *Roller-Coaster: Europe, 1950-2017*

Adam Tooze, *Crashed: How a Decade of Financial Crises Changed the World*

Edmund King, *Henry I: The Father of His People*

Lilia M. Schwarcz and Heloisa M. Starling, *Brazil: A Biography*

Jesse Norman, *Adam Smith: What He Thought, and Why it Matters*

Philip Augur, *The Bank that Lived a Little: Barclays in the Age of the Very Free Market*

Christopher Andrew, *The Secret World: A History of Intelligence*

David Edgerton, *The Rise and Fall of the British Nation: A Twentieth-Century History*

Julian Jackson, *A Certain Idea of France: The Life of Charles de Gaulle*

Owen Hatherley, *Trans-Europe Express*

Richard Wilkinson and Kate Pickett, *The Inner Level: How More Equal Societies Reduce Stress, Restore Sanity and Improve Everyone's Wellbeing*

Paul Kildea, *Chopin's Piano: A Journey Through Romanticism*

Seymour M. Hersh, *Reporter: A Memoir*

Michael Pollan, *How to Change Your Mind: The New Science of Psychedelics*

David Christian, *Origin Story: A Big History of Everything*

Judea Pearl and Dana Mackenzie, *The Book of Why: The New Science of Cause and Effect*

David Graeber, *Bullshit Jobs: A Theory*

Serhii Plokhy, *Chernobyl: History of a Tragedy*

Michael McFaul, *From Cold War to Hot Peace: The Inside Story of Russia and America*

Paul Broks, *The Darker the Night, the Brighter the Stars: A Neuropsychologist's Odyssey*

Lawrence Wright, *God Save Texas: A Journey into the Future of America*

John Gray, *Seven Types of Atheism*

Carlo Rovelli, *The Order of Time*

Mariana Mazzucato, *The Value of Everything: Making and Taking in the Global Economy*

Richard Vinen, *The Long '68: Radical Protest and Its Enemies*

Kishore Mahbubani, *Has the West Lost It?: A Provocation*

John Lewis Gaddis, *On Grand Strategy*

Richard Overy, *The Birth of the RAF, 1918: The World's First Air Force*

Francis Pryor, *Paths to the Past: Encounters with Britain's Hidden Landscapes*

Helen Castor, *Elizabeth I: A Study in Insecurity*

Ken Robinson and Lou Aronica, *You, Your Child and School*

Leonard Mlodinow, *Elastic: Flexible Thinking in a Constantly Changing World*

Nick Chater, *The Mind is Flat: The Illusion of Mental Depth and The Improvised Mind*

Michio Kaku, *The Future of Humanity: Terraforming Mars, Interstellar Travel, Immortality, and Our Destiny Beyond*

Thomas Asbridge, *Richard I: The Crusader King*

Richard Sennett, *Building and Dwelling: Ethics for the City*

Nassim Nicholas Taleb, *Skin in the Game: Hidden Asymmetries in Daily Life*

Steven Pinker, *Enlightenment Now: The Case for Reason, Science, Humanism and Progress*

Steve Coll, *Directorate S: The C.I.A. and America's Secret Wars in Afghanistan, 2001 - 2006*

Jordan B. Peterson, *12 Rules for Life: An Antidote to Chaos*

Bruno Maçães, *The Dawn of Eurasia: On the Trail of the New World Order*

Brock Bastian, *The Other Side of Happiness: Embracing a More Fearless Approach to Living*

Ryan Lavelle, *Cnut: The North Sea King*

Tim Blanning, *George I: The Lucky King*

Thomas Cogswell, *James I: The Phoenix King*

Pete Souza, *Obama, An Intimate Portrait: The Historic Presidency in Photographs*

Robert Dallek, *Franklin D. Roosevelt: A Political Life*

Norman Davies, *Beneath Another Sky: A Global Journey into History*

Ian Black, *Enemies and Neighbours: Arabs and Jews in Palestine and Israel, 1917-2017*

Martin Goodman, *A History of Judaism*

Shami Chakrabarti, *Of Women: In the 21st Century*

Stephen Kotkin, *Stalin, Vol. II: Waiting for Hitler, 1928-1941*

Lindsey Fitzharris, *The Butchering Art: Joseph Lister's Quest to Transform the Grisly World of Victorian Medicine*

Serhii Plokhy, *Lost Kingdom: A History of Russian Nationalism from Ivan the Great to Vladimir Putin*

Mark Mazower, *What You Did Not Tell: A Russian Past and the Journey Home*

Lawrence Freedman, *The Future of War: A History*

Niall Ferguson, *The Square and the Tower: Networks, Hierarchies and the Struggle for Global Power*

Matthew Walker, *Why We Sleep: The New Science of Sleep and Dreams*

Edward O. Wilson, *The Origins of Creativity*

John Bradshaw, *The Animals Among Us: The New Science of Anthropology*

David Cannadine, *Victorious Century: The United Kingdom, 1800-1906*

Leonard Susskind and Art Friedman, *Special Relativity and Classical Field Theory*

Maria Alyokhina, *Riot Days*

Oona A. Hathaway and Scott J. Shapiro, *The Internationalists: And Their Plan to Outlaw War*

Chris Renwick, *Bread for All: The Origins of the Welfare State*

Anne Applebaum, *Red Famine: Stalin's War on Ukraine*

Richard McGregor, *Asia's Reckoning: The Struggle for Global Dominance*

Chris Kraus, *After Kathy Acker: A Biography*

Clair Wills, *Lovers and Strangers: An Immigrant History of Post-War Britain*

Odd Arne Westad, *The Cold War: A World History*

Max Tegmark, *Life 3.0: Being Human in the Age of Artificial Intelligence*

Jonathan Losos, *Improbable Destinies: How Predictable is Evolution?*

Chris D. Thomas, *Inheritors of the Earth: How Nature Is Thriving in an Age of Extinction*

Chris Patten, *First Confession: A Sort of Memoir*

James Delbourgo, *Collecting the World: The Life and Curiosity of Hans Sloane*

Naomi Klein, *No Is Not Enough: Defeating the New Shock Politics*

Ulrich Raulff, *Farewell to the Horse: The Final Century of Our Relationship*

Slavoj Žižek, *The Courage of Hopelessness: Chronicles of a Year of Acting Dangerously*

Patricia Lockwood, *Priestdaddy: A Memoir*

Ian Johnson, *The Souls of China: The Return of Religion After Mao*

Stephen Alford, *London's Triumph: Merchant Adventurers and the Tudor City*

Hugo Mercier and Dan Sperber, *The Enigma of Reason: A New Theory of Human Understanding*

Stuart Hall, *Familiar Stranger: A Life Between Two Islands*

Allen Ginsberg, *The Best Minds of My Generation: A Literary History of the Beats*

Sayeeda Warsi, *The Enemy Within: A Tale of Muslim Britain*

Alexander Betts and Paul Collier, *Refuge: Transforming a Broken Refugee System*

Robert Bickers, *Out of China: How the Chinese Ended the Era of Western Domination*

Erica Benner, *Be Like the Fox: Machiavelli's Lifelong Quest for Freedom*

William D. Cohan, *Why Wall Street Matters*

David Horspool, *Oliver Cromwell: The Protector*

Daniel C. Dennett, *From Bacteria to Bach and Back: The Evolution of Minds*

Derek Thompson, *Hit Makers: How Things Become Popular*

Harriet Harman, *A Woman's Work*

Wendell Berry, *The World-Ending Fire: The Essential Wendell Berry*

Daniel Levin, *Nothing but a Circus: Misadventures among the Powerful*

Stephen Church, *Henry III: A Simple and God-Fearing King*

Pankaj Mishra, *Age of Anger: A History of the Present*

Graeme Wood, *The Way of the Strangers: Encounters with the Islamic State*

Michael Lewis, *The Undoing Project: A Friendship that Changed the World*

John Romer, *A History of Ancient Egypt, Volume 2: From the Great Pyramid to the Fall of the Middle Kingdom*

Andy King, *Edward I: A New King Arthur?*

Thomas L. Friedman, *Thank You for Being Late: An Optimist's Guide to Thriving in the Age of Accelerations*

John Edwards, *Mary I: The Daughter of Time*

Grayson Perry, *The Descent of Man*

Deyan Sudjic, *The Language of Cities*

Norman Ohler, *Blitzed: Drugs in Nazi Germany*

Carlo Rovelli, *Reality Is Not What It Seems: The Journey to Quantum Gravity*

Catherine Merridale, *Lenin on the Train*

Susan Greenfield, *A Day in the Life of the Brain: The Neuroscience of Consciousness from Dawn Till Dusk*

Christopher Given-Wilson, *Edward II: The Terrors of Kingship*

Emma Jane Kirby, *The Optician of Lampedusa*

Minoo Dinshaw, *Outlandish Knight: The Byzantine Life of Steven Runciman*

Candice Millard, *Hero of the Empire: The Making of Winston Churchill*

Christopher de Hamel, *Meetings with Remarkable Manuscripts*

Brian Cox and Jeff Forshaw, *Universal: A Guide to the Cosmos*

Ryan Avent, *The Wealth of Humans: Work and Its Absence in the Twenty-first Century*

Jodie Archer and Matthew L. Jockers, *The Bestseller Code*

Cathy O'Neil, *Weapons of Math Destruction: How Big Data Increases Inequality and Threatens Democracy*

Peter Wadhams, *A Farewell to Ice: A Report from the Arctic*

Richard J. Evans, *The Pursuit of Power: Europe, 1815-1914*

Anthony Gottlieb, *The Dream of Enlightenment: The Rise of Modern Philosophy*

Marc Morris, *William I: England's Conqueror*

Gareth Stedman Jones, *Karl Marx: Greatness and Illusion*

J.C.H. King, *Blood and Land: The Story of Native North America*

Robert Gerwarth, *The Vanquished: Why the First World War Failed to End, 1917-1923*

Joseph Stiglitz, *The Euro: And Its Threat to Europe*

John Bradshaw and Sarah Ellis, *The Trainable Cat: How to Make Life Happier for You and Your Cat*

A J Pollard, *Edward IV: The Summer King*

Erri de Luca, *The Day Before Happiness*